W9-BRU-343

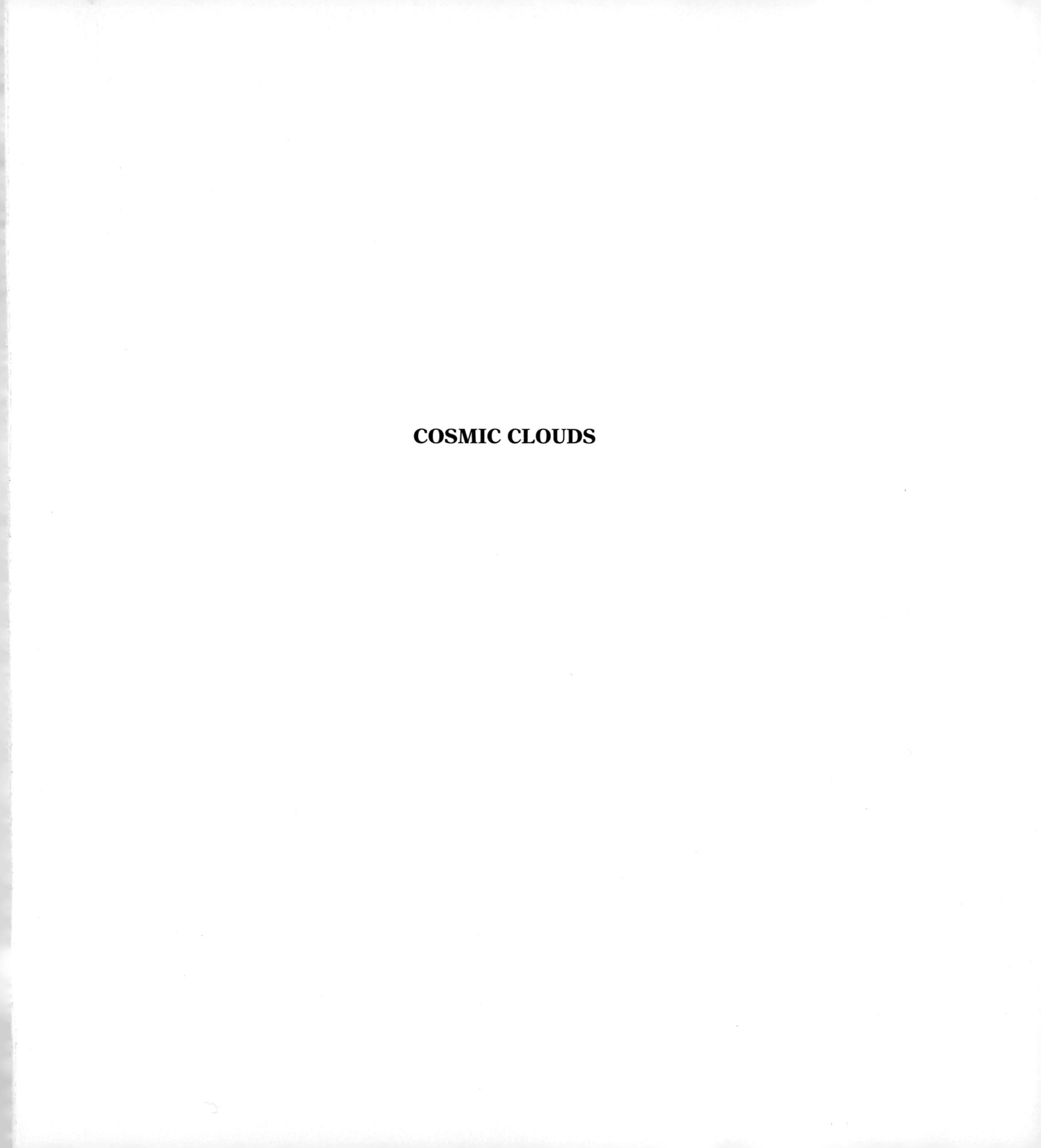

COSMIC CLOUDS

< The vast spaces between the stars are dynamic and full of life. We see black, dusty globules and huge, dark molecular clouds giving birth to stars that in turn illuminate their surroundings, old stars exhaling matter back into space, and massive exploding stars violently heating and stirring the interstellar gas—all the activity parts of cycles of stellar birth, death, and regeneration.

COSMIC CLOUDS

Birth, Death, and Recycling in the Galaxy

James B. Kaler

SCIENTIFIC
AMERICAN
LIBRARY

A Division of HPHLP
New York

Cover and Text Designer: Diana Blume

Library of Congress Cataloging–in–Publication Data

Kaler, James B.
Cosmic clouds : birth, death, and recycling in the galaxy / James B. Kaler
p. cm.
Includes bibliographical references and index.
ISBN 0-7167-5075-9
1. Galaxies. 2. Stars. 3. Astrophysics. I. Title.
QB857.K35 1997 96-46769
523.1'135—dc21 CIP

The story of Black God and the stars, which appears at the beginning of Chapter 6, is told in Ray A. Williamson, *Living the Sky: The Cosmos of the American Indian* (Boston: Houghton Mifflin Company, 1984).

ISSN 1040-3213

Printed in the United States of America

Scientific American Library
A Division of HPHLP
New York

Distributed by W. H. Freeman and Company
41 Madison Avenue, New York, NY 10010
Houndmills, Basingstoke RG21 6XS, England

First printing 1997

This book is number 64 of a series.

To the memory of my grandmother, Laura Jaekel Kaler,
who started me on this exciting road,
and to my parents, Hazel Holmgren (Susie) Kaler and Earl Kaler,
who helped keep me there

Contents

PREFACE

Cosmic Clouds is designed to place the stars, our Earth, and ourselves into the broad context of the Galaxy. Here we explore the vast clouds of interstellar space and how they condense into the stars (including our Sun) that shine around us; how star birth generates planets and perhaps life itself; and how the deaths of stars provide pathways, other interstellar clouds, that return us to the vastness of space—the whole process an exercise in cosmic recycling, new stars born from the debris of the old. A companion to *Stars* (also published by Scientific American Library), *Cosmic Clouds* looks at the stars from a different perspective, from the outside rather than from the inside.

Science books mark moments in time. While they are being produced, the science relentlessly continues onward. The subjects of star birth, planet formation, and interstellar recycling are dynamic ones, in states of constant change: planets orbiting other stars, for example, are now being discovered at the rate of about one a month. Much of the foundation of this broad subject, however, is solidly built and will withstand further investigation. *Cosmic Clouds* is therefore an amalgam of thoroughly understood science and science at the exhilarating edge of discovery, a combination that provides the reader with a base from which to explore the future.

The concepts developed herein, and the linkages of these moments in time, have gone through several revisions that would not have been possible without the excellent work of those at Scientific American Library. I would like to thank senior editor Jonathan Cobb for his interest and insight when approached with the initial idea for the book; his commitment to the project made it possible. Deep and appreciative thanks go to my technical readers and advisors: Bruce Elmegreen of the Watson Research Center of IBM, John Graham of the Department of Terrestrial Magnetism, Carnegie Institution of Washington, Eugene Levy of the University of Arizona, and Robert O'Dell of Rice University, who all read the entire manuscript; Virginia Trimble of the University of California at Irvine, who critiqued the first few chapters; Scott Kenyon of the Center for Astrophysics at Harvard, who instructed me on the subtleties of star formation; and my Illinois colleagues John Dickel, Icko Iben, Margaret Meixner, Telemachos Mouschovias, Lew Snyder, and Ken Yoss, who patiently answered my many questions.

Thanks, too, to the many people who graciously gave their time to provide many of the book's images.

Cosmic Clouds was guided through its own dynamic process of production by my old (then unknown) upstate neighbor Mary Louise Byrd. It is filled with images found with the help of talented photo editors Barbara Brooks and Larry Marcus; Tomo Narashima created the lively cosmos seen in the frontispiece. All these elements have been brought together with the text by the sensitive work of designer Diana Blume. A special place is reserved for my friend and editor Nancy Brooks, who, with Jonathan Cobb, helped develop the conceptual structure of the book, who critiqued and improved the line art, and who line by line helped link my words into a fluid whole, providing an example of thoughtful editing at its finest. And finally, thanks, as always, to my wife, Maxine, for her continuing support.

COSMIC CLOUDS

1

Among the Stars

Stars of the Milky Way crowd together, seeming almost to touch.

The rolling Earth turns away from the Sun and the continuous blue of the day into the shadow of night, the darkened sky becoming punctuated with thousands of stars. They seem to be everywhere, dominating the heavens with their light. Photography reinforces this view: in dense portions of the Milky Way, stellar images appear to overlap, giving the effect of a near-continuous sheet of light. Stars appear in such mutual proximity it seems a mystery how they avoid crashing together.

The effect is a grand illusion. The image of a star seen through the telescope is a disk produced by the smearing effects of the optics and our terrestrial atmosphere: it appears to have dimension. In reality, however, the stars are so far away that no matter how physically large they are, they effectively appear as points. Even those nearest Earth have angular dimensions typically only a thousandth of a second of arc across, the angular size of a U.S. penny seen from 3000 km away. If they were not blurred, the star images in the chapter-opening photograph would be a mere 0.00001 mm wide. Factoring in the third dimension—the enormous differences in their distances from Earth—stars are typically separated from each other by over 10 million times their diameters. The nighttime sky—and space itself—may appear to be filled with stars, but in fact it is remarkably empty of them: of the volume of space, only 1 part in 10^{21} is actually filled with a star.

What lies in the vast room between the stars? More than could be imagined even as recently as 30 years ago. Over the past 200 years, with discoveries proceeding at a wildly accelerating pace during the latter part of our own century, astronomers have found colorful puffs of glowing gas, glowering black billows of obscuring dust filled with intricate molecular chemistries, graceful luminous winds of gas that blow from dying stars, violent expanding blast waves from exploded stars, and much more.

The cosmos, like the sky on a turbulent summer day, is filled with clouds of different sizes, shapes, structures, and distances; some are thick, swelling cumulus, others, light wispy cirrus—all of them constantly changing, colliding, forming, and evaporating. Some of summer's clouds thicken enough to condense and produce raindrops that ultimately evaporate away into the skies to generate more clouds. The night's stars are the raindrops of the interstellar clouds, and like water droplets, they—or at least part of them—recycle, evaporating back to feed the interstellar medium, the fragmented, chaotic, dusty gases of interstellar space from which they came.

The rain washes the air, carrying the wind-blown dust of the sky back to the ground. The planets, including our own Earth, were made from the dusty leavings of the matter from which our own raindrop, our Sun, was created. Each atom of our world, each atom of ourselves, was once in interstellar space. The Earth is a distillate of the interstellar medium and therefore is also a by-product of earlier stellar generations. We are not at all separate from the night's stars, but through the churning creative processes of the interstellar medium are very much a part of them.

THE STELLAR CONTEXT

Though the interstellar medium is widely dispersed, it still in a sense belongs to the stars. It is a way station in stellar evolution, providing the birthplace, the recycling bin, and the burial ground of stars. We therefore begin to examine it by looking at the stars and the context they provide.

Where and what are the stars? How do we know? How do they relate to this enigmatic interstellar medium that apparently created them? We have a fine example of condensed interstellar matter quite close to us, the Sun. The Sun (like many other stars, as we are beginning to learn) has a family. Orbiting it are eight major bodies, the planets, and a huge variety of smaller stuff. In his great book, *De Revolutionibus,* published in 1543, Nicolas Copernicus gave the relative distances from the Sun of the five other planets known at the time, Mercury through Saturn, in terms of the distance between the Earth and Sun. This distance, our first astronomical yardstick, is so important it is now known simply as the astronomical unit, or AU. Somewhat over a century later, Giovanni Cassini found the distance to Mars from its parallax, its shift in angular position when viewed from different points on Earth. The measurement of the Martian distance provided a means to determine the length of the AU; the modern value is 150 million km.

With the Sun's distance known, we can calculate its diameter—1.5 million km, 109 times that of Earth. Shortly after Copernicus forever displaced the Earth from the center of the Universe, Johannes Kepler, working from the observations of Tycho Brahe, the last and greatest of the naked-eye observers, laid down the laws for planetary orbits. He found that they are ellipses and, most important, that their orbital periods and dimensions are related: the square of the period (P) in years always equals the cube of the ellipse's semimajor axis (a, half the length of the longest line that can be drawn in the ellipse), expressed in AU (or $P^2_{\text{yr}} = a^3_{\text{AU}}$). In 1687, Isaac Newton (through his laws of motion and the law of gravity) reformulated Kepler's laws in an elegant quantitative form that links

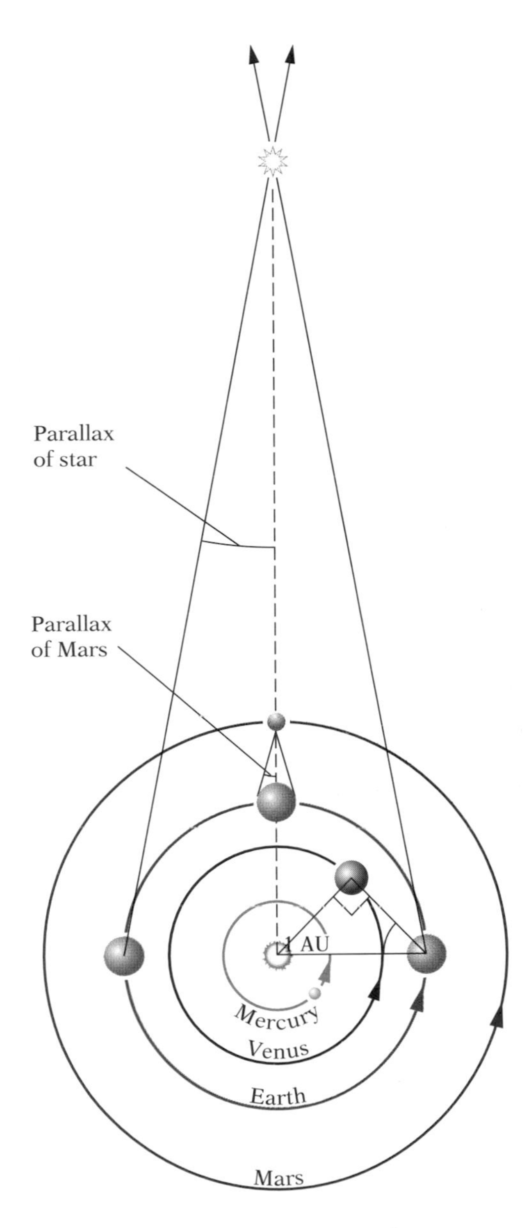

The maximum angle between a planet (here, Venus) and the Sun allows the determination of the distance of the planet in terms of the AU. The AU can be calibrated from the distance of any body, found by parallax (as here), or by radar. The annual shifts of the nearer stars as the Earth revolves in orbit allow the determination of the stellar distances in AU and thus in kilometers.

the period, semimajor axis, and the sum of the masses of *any* pair of orbiting bodies in *physical* units, respectively, seconds, centimeters, and grams (P^2 = constant $\times$ $a^3/(M_1 + M_2)$). (Kepler's original law avoids the masses because the mass of the Sun totally dominates the masses of the individual planets.) From the orbital characteristics of the Earth, astronomers were able to determine the combined masses of the Sun and Earth and, since the Earth's mass is comparatively inconsequential, of the mass of the Sun alone: the Sun contains the mass of 330,000 Earths. From this huge body pours an energy flow equal to 4 $\times$ 10^{26} watts. Concentrated interstellar matter is potent indeed.

Without any real proof, the Greek astronomer Aristarchus, in the third century B.C., believed the stars to be vastly farther away than the Sun (which from the timing of lunar phases he determined to be 20 times more distant than the Moon, a result too short by a factor of 20). Just how distant the stars are was not known until surprisingly recent times. If Copernicus was right, the stars should show parallaxes. As the Earth orbits the Sun, we can observe the stars from two different vantage points in space. Yet no one could detect these parallaxes. If the angle of parallax was that small, the stars were truly immensely distant. Not until 1846 did the German astronomer Friedrich Bessel succeed in observing the parallax of the nearby star 61 Cygni: only two-thirds of a second of arc, the diameter of a U.S. nickel seen at a distance of 6 km. At last we knew: 61 Cygni is over 600,000 AU away. Such numbers require the use of larger units. The light-year, the distance light travels in a year (at 300,000 km/s), is 63,271 AU long. Another unit, the parsec (*par*allax + *sec*ond), is the distance (206,265 AU, or 3.26 ly) at which the radius of the Earth's orbit would be 1 second of arc across. The star 61 Cygni is thus about 10 ly, or 3 pc, away. Modern techniques allow the measurement of tens of thousands of stellar parallaxes to distances of hundreds of parsecs. Other distance methods calibrated by the parallax procedure allow us to measure much of the Universe.

The distances and apparent brightnesses of the stars reveal that they range in luminosity—that is, in total radiant energy output per second—from a million times that of the Sun to a million times less. If the brightest of them were at the distance of 61 Cygni, it would shine almost as brightly in our sky as the full Moon. The dimmest stars would have to be very close, less than a hundred times the distance of the planet Pluto (only 40 AU away), just to be seen. The processes that produce stars from the interstellar medium—and the medium itself—must be enormously complex to create some stars a trillion times brighter than others.

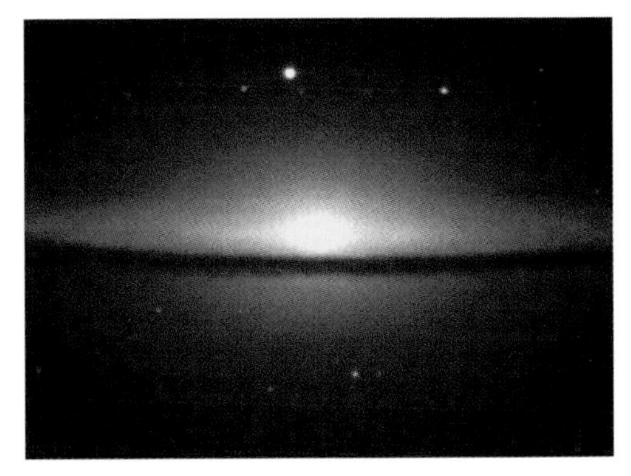

To the left, a nearby galaxy called M 51 at a glance shows a typical arrangement of stars, here seen in a face-on disk with graceful spiral arms. Ours might look something like this one if seen from a great distance; the Sun would be about two-thirds of the way from the center to the apparent edge. Above, the Sombrero Galaxy (M 104) shows its nearly edge-on disk and its huge central bulge, which extends outward into a great halo filled with giant globular clusters of stars.

Stars cannot exist, or be created, in isolation. Some kind of overall organization must arrange and compress the interstellar medium. Not until the 1920s did astronomers confirm that all the stars of the Universe are organized into billions of individual collections called galaxies that differ greatly from one another in size and structure. Almost no stars lie between the galaxies. Ours, parochially called simply "the Galaxy," contains all the stars you see at night and some 200 billion others. Its principal feature, a flat disk that contains over 95 percent of stellar collection, is about 35,000 pc (100,000 ly) across. Enclosing the disk is a thinly populated, spheroidal halo. Around the galactic center, the disk thickens into a large bulge about 2000 pc wide.

Our Galaxy's disk encircles us as the Milky Way. From our off-center position toward the Galaxy's edge, we look back to see the vast star clouds toward the center, the disk —cut by clouds of dark interstellar dust— fading away to both sides.

The stars of the Galaxy are bound together under the force of their mutual gravity. If they were stationary, gravity would cause them all to fall together at the center in a heap, and we could not exist. The Galaxy must therefore be a dynamic entity. Just as the planets orbit the Sun, all the Galaxy's stars must orbit the Galaxy's center. In the simplest sense, each star feels the combined force of all the other stars inside its orbit collectively acting as its gravitational "sun," and each star orbits under their mutual influence. Since the stars of the disk move in the same direction on relatively circular orbits, the Galaxy, as a whole, is said to rotate.

The Galaxy presents itself to us with grace and beauty. Stand under a dark sky on an evening in northern summer or autumn (winter or spring in the southern hemisphere). You will see, arching across the heavens from one horizon to the other, a wide glowing band of light, the Milky Way. Known by different names in the world's cultures, it is the source of innumerable mythologies and legends. When Galileo examined it in 1609 with the first astronomical telescope, he saw it resolved into countless stars: it is our Galaxy's disk. Our Sun resides within the disk, about 8000 pc out from the Galaxy's core, and therefore we see the combined light of the disk's stars as an encircling belt. In part because of our off-center position, the visibility of the Milky Way varies considerably. In the northern summer, observers see the Milky Way coming out of the northeast through the constellation Cassiopeia, then overhead through Cygnus, where it appears to break in two. Toward the south, the western branch flows to the right, into Ophiuchus and Scorpius, where it terminates. The eastern branch continues to brighten through the star clouds of Scutum and then gloriously into Sagittarius. Here is the heart of the Milky Way, the fields of stars thickening in the direction of the center of the Galaxy, toward which Sagittarius unknowingly points his mythical arrow.

The galactic disk and halo are profoundly different from each other. As well as containing most of the Galaxy's stars, the disk includes by far the bulk of the interstellar medium, showing us that star birth has ceased in the halo and that it can contain only old stars. On a moonless night deep in the countryside, you can see with your eye alone that the Milky Way is enormously irregular, varying mightily in width and brightness. Throughout there are patches of black, small and large, that appear to be holes cut through the Galaxy itself. With the telescope, you can also see faint glowing gas clouds that are often allied with the gaps. The gaps are not voids in the stellar distribution but are dense clouds of interstellar dust that block the light of the background, clouds that are creating new stars that will someday join their uncountable older siblings that lie within the milky circle. The bright clouds are the dark clouds' tattered remnants, the brighter stars forged within both illuminating the clouds and tearing them apart, the shredded matter re-forming to produce yet more stellar generations.

We therefore cannot understand the stars—and their planets—without first knowing the interstellar medium, both as a separate entity and as a vital part of the Galaxy as a whole. The story of how astronomers finally learned the nature of the Galaxy's structure is in fact intertwined with the discovery of the interstellar medium and with the discrimination among the various types of interstellar clouds, galaxies, star clusters, and other celestial objects.

THE RIDDLE OF THE NEBULAE

With the invention of the telescope in the early 1600s astronomers began to find amorphous patches of light apparently floating amidst the stars. Some of them moved briefly against the stellar background,

227 1784.

CATALOGUE

DES

NÉBULEUSES ET DES AMAS D'ÉTOILES

Obſervées à Paris, par M. Meſſier, à l'Obſervatoire de la Marine, hôtel de Clugni, rue des Mathurins.

M. MESSIER a obſervé avec le plus grand ſoin les Nébuleuſes & les amas d'Étoiles qu'on découvre ſur l'horizon de Paris; il a déterminé leur aſcenſion droite, leur déclinaiſon, & donné leurs diamètres, avec des détails circonſtanciés ſur chacune : ouvrage qui manquoit à l'Aſtronomie.

Il entre auſſi dans des détails ſur les recherches qu'il a faites des différentes Nébuleuſes qui ont dû être découvertes par différens Aſtronomes, mais qu'il a cherchées inutilement.

Le Catalogue des Nébuleuſes & des amas d'Étoiles, de M. Meſſier, eſt inſéré dans le volume de l'Académie des Sciences, *année 1771, page 435.* Il rapporte à la fin de ſon Mémoire, un deſſin tracé avec le plus grand ſoin de la belle Nébuleuſe de l'épée d'Orion, avec les étoiles qu'elle contient. Ce deſſin pourra ſervir à reconnoître ſi dans la ſuite des temps elle n'eſt pas ſujette à quelque changement. Si l'on compare dès-à-préſent ce deſſin avec ceux de Mrs Huyghens, Picard, de Mairan & le Gentil, on ſera étonné d'y trouver un changement tel qu'on auroit peine à ſe figurer que c'eſt la même nébuleuſe, ſi l'on n'avoit égard qu'à ſa figure. On peut voir ces deſſins, donnés par M. le Gentil dans le volume de l'Académie, de 1759, *page 470, planche XXI.*

Au Catalogue imprimé de M. Meſſier, que nous donnons ici, nous rapporterons encore un grand nombre de nébuleuſes & d'amas d'étoiles qu'il a découvertes depuis l'impreſſion de ſon Mémoire, & qu'il nous a communiquées.

Aux poſitions des nébuleuſes, M. Meſſier a rapporté des numéros qui ſont les mêmes à la page ſuivante, & qui donnent le détail de chacune des nébuleuſes obſervées.

The frontispiece of Messier's 1784 Catalogue of Nebulous Objects and Star Clusters.

The Double Cluster in Perseus (above) is a pair of common open star clusters (named for their loose, open appearance) that to the naked eye look like fuzzy balls. The great globular cluster Omega Centauri (on the facing page) contains over a million stars.

then disappeared. These were readily recognized as comets, which have been observed with the naked eye throughout human history and are now known to be tiny icy bodies that inhabit the outer Solar System. Other luminous celestial blobs, however, held still among the stars. To avoid confusion between them and the comets, in the 1780s the French astronomer Charles Messier produced the first definitive catalogue of the brightest of the stationary objects, including some like the Pleiades (M 45) that were known to be clusters of stars. His initial tabulation of 103 objects, later extended to 109, includes some of the brightest and loveliest of the sky's "non-stellar objects" (still so called even if they are composed of stars) and is thus a centerpiece of amateur observing.

Over the next 20 years, William Herschel added another 2500 to the list of fuzzy objects, cataloguing them as the Earth's rotation swept them through his telescope's field of view. Herschel, a musician, was born in Hannover in 1738. He left the military bands of the

Hannoverian army in 1756 for England, where he taught, performed, and composed. His real fascination, and greatest skills, lay in astronomy, and by 1774 he was scanning the skies with a reflecting telescope of his own making and beginning to construct ever larger and better instruments. His work, for which he was knighted in 1816, includes the discoveries of the planet Uranus, double stars, and infrared radiation. His son, John, who extended his father's scientific inquiries into the southern hemisphere, was knighted in his turn by Queen Victoria on the occasion of her coronation. The Herschels' *General Catalogue* of non-stellar objects was updated in 1888 by J. L. E. Dryer as the *New General Catalogue* (*NGC*). This still-standard work, which lists over 7000 non-stellar objects, was extended again by the *Index Catalogue* (*IC*), which contains another 5000.

Many of these catalogued objects were seen simply to be more star clusters so far away that through poorer instruments the images blended together into luminous lumps. Such clusters were everywhere, and as telescopes improved, they were discovered to be of two distinct kinds. Loose "open" clusters, exemplified by the Pleiades, the Beehive, the Double Cluster, and several others visible to the naked eye, throng the Galaxy's disk. Open clusters are mostly small, ragged systems of a few hundred stars, each stuffed within a volume 10 to 20 pc across. Two hundred or more massive "globular" clusters, some of which contain a million or more stars, occupy the galactic halo. The Milky Way's interstellar dust is so thick that we cannot see very far within the disk, but the globulars, outside the disk, are easily visible. Employing methods calibrated by parallaxes, astronomers can find their distances from examination of the stars they contain. From such measurements the distribution of the globular clusters can be found. If we assume that the center of their distribution coincides with the center of the Galaxy, we can find the Galaxy's dimensions.

Other non-stellar objects defied resolution into stars and were termed nebulae (the Latin word for "clouds"); among them were the "great nebulae" in Andromeda and Orion. The Andromeda Nebula (M 31), an elongated patch roughly a degree in length that is rather easily visible to the naked eye, was mentioned by the tenth-century Arabian astronomer Al-Sufi. In the first telescopic observation in 1612, the German astronomer Simon Marius likened it to a candle shining through a translucent piece of horn. The Orion Nebula (M 42) is a vast swirling, complex cloud of very roughly the same angular extent, with a quartet of obvious bright stars, the Trapezium, at its center; it was found by Nicolas de Peiresc in 1611.

These nebulae and others like them were clearly not clusters. They remained unresolved into anything, appearing only as glowing clouds in spite of the increasingly great telescopic power applied. Opinion as to their natures was sharply divided. William Parsons, the third earl of Rosse, had by 1845 built the largest telescope in the world, a reflector 72 inches in aperture, at Birr Castle in Ireland. With it, he discovered that one of the mysterious nebulae, M 51 in Canes Venatici (at the end of the handle of the Big Dipper), had a spiral structure. As telescopes improved, more and more non-stellar objects were revealed as clusters of stars, and Parsons thus felt that all the nebulae, including his new class of "spiral nebulae," would also eventually be seen to be made of stars. In 1847, Harvard's G. P. Bond found the spiral arms in M 31, which thus suggested a stellar composition. He also claimed to have seen the brightest portions of the Orion Nebula as masses of tiny stars and made a case that it too possessed a spiral structure, putting it in the same class with the Andromeda Nebula.

Other kinds of objects, however, cast doubt on the stellar theory. Among William Herschel's discoveries were a number of small glowing clouds he called "planetary nebulae" because of their disklike shapes. Through the telescope they have somewhat the same soft appearance as the Orion Nebula but are much smaller; like the Orion Nebula (and unlike the Andromeda Nebula), they have stars at their centers. Herschel was convinced, because of the "milkiness or soft tint of their light," that the planetary nebulae were not composed of stars but consisted of "a shining fluid of a nature totally unknown to us." By implication, so did the Orion Nebula and several other similar objects. In 1785, Herschel had suggested that the planetary nebulae might be stars caught in the act of "regeneration." A few years later he surmised that the obvious stars at the center of the Orion Nebula and those bright ones scattered throughout might be attracting the surrounding fluid, eventually to appear as planetary nebulae, and that the fluid "seems . . . fit to produce a

M 31, now known to be a large, nearby galaxy, requires a large telescope to resolve it into stars and in that way discriminate it from gaseous clouds.

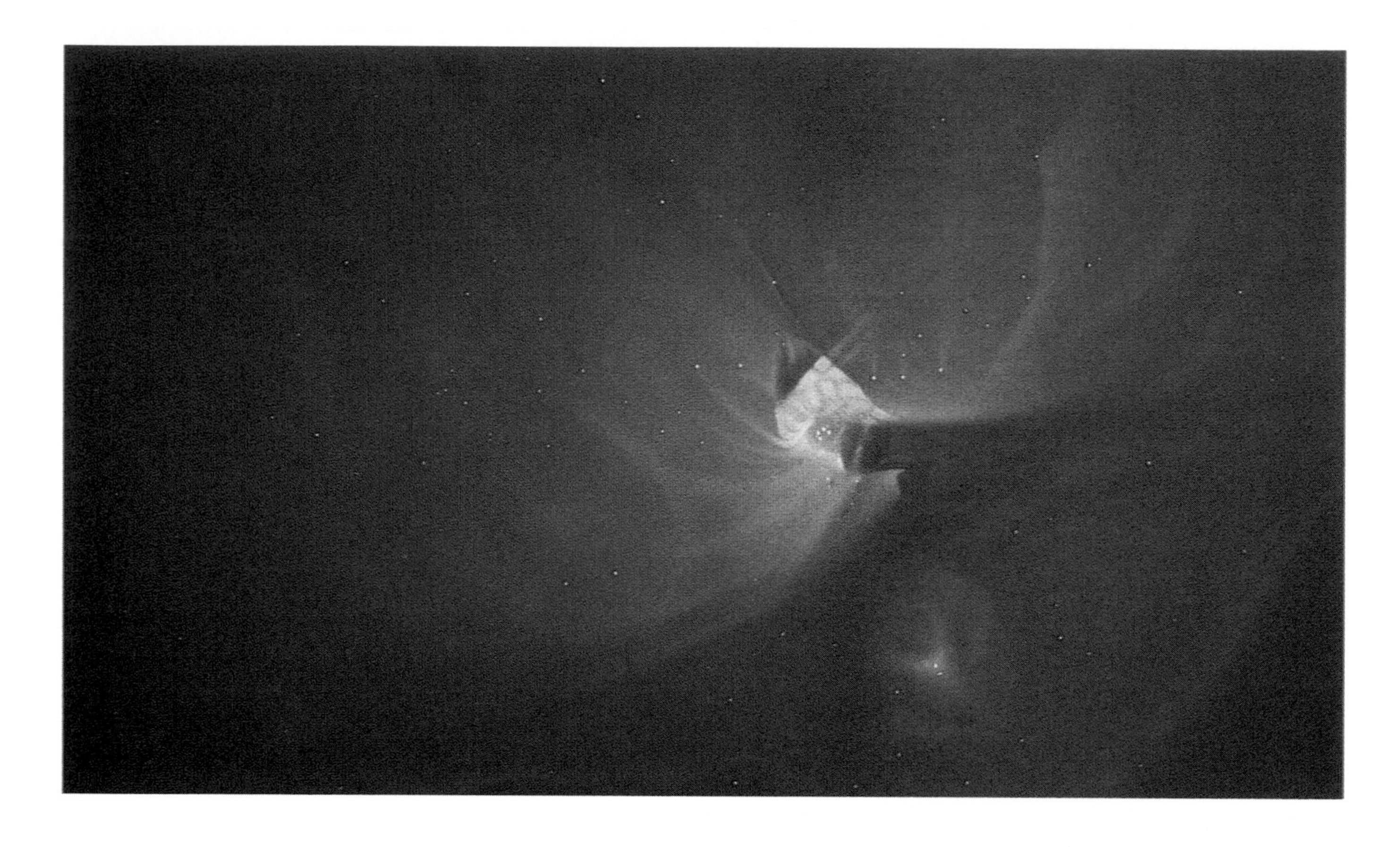

G. P. Bond's exquisite drawing of the Orion Nebula, a true gas cloud, shows what he thought to be masses of unresolved stars.

star by its condensation." Though erroneous in concept, this was the first serious hint of the ability of the interstellar medium to create stars.

Confusion reigned and arguments raged, arguments not unlike those involving other kinds of celestial objects today. Astronomers saw spiral nebulae that were believed to be composed of unresolved stars, Orion-like nebulae that might or might not be made of unresolved stars, and planetary nebulae that seemed to be something different but might be related to the Orion-like objects. How were astronomers ever to find out what any of these "nebulae" were? The key that unlocked the secrets—to the nebulae, to the nature of our Galaxy, to the existence of other galaxies, and eventually to star formation—was to be found in the light of the celestial bodies themselves. When in 1835 the French philosopher Auguste Comte forcefully asserted that "never, by any means, will we be able to study their chemical composition, their mineralogic structure" and that "men will never encompass in their conceptions the whole of the stars," he had not reckoned with the power of the rainbow.

Reading the Rainbows

A hundred and fifty years before Comte's time, Isaac Newton had placed a glass prism in a shaft of sunlight and watched the beam break into a dance of dazzling color, a rainbow with deep red at one end and violet at the other. We now know that light can be thought of as either a traveling wave of alternating electric and magnetic fields or as a particle, a *photon,* that carries energy, and in fact it is both. The phenomenon we perceive as color depends on the photon's wavelength or frequency. Photons of violet light have wavelengths of some 4×10^{-5} cm (4000 Å, where an angstrom, Å, is 10^{-8} cm), those of red light about 7000 Å. This array, or *spectrum,* of light extends to shorter, invisible waves, into the ultraviolet, X rays, and gamma rays, and to longer ones, into the infrared and radio domains. A photon's energy is directly proportional to frequency (hence inversely proportional to wavelength). A gamma-ray photon, with a wavelength of less than an angstrom, carries thousands of times the impact of an optical photon (one detectable by the eye) and billions of times more than a radio photon. Low-energy radio waves are benign; gamma rays, produced copiously by atomic bomb explosions, are deadly.

Look into the nighttime sky. The brighter stars have different (albeit subtle) colors that range from reddish, through yellow-white like sunlight, and on to those with a pale blue tint. The color of a star is an index of its temperature, a correlation possible because stars are reasonable approximations to an idealization of nature called a blackbody, a body defined as one that absorbs all the radiation that falls upon it. To maintain a blackbody at a constant temperature, it must also emit radiation and, like the Sun, can perversely be very bright. The temperature of a body, be it gas or solid, bowling ball or star, is a measure of its internal energy, which is reflected in the speeds (vibration speeds in a solid) of its atoms, the particles of which it is composed. Remove some energy and the body cools; remove it all and the temperature plunges to absolute zero, –273°C, or 0 degrees Kelvin (the Kelvin scale, K, counts upward from zero in centigrade degrees). A blackbody in space must radiate energy. A low-temperature body is capable of producing

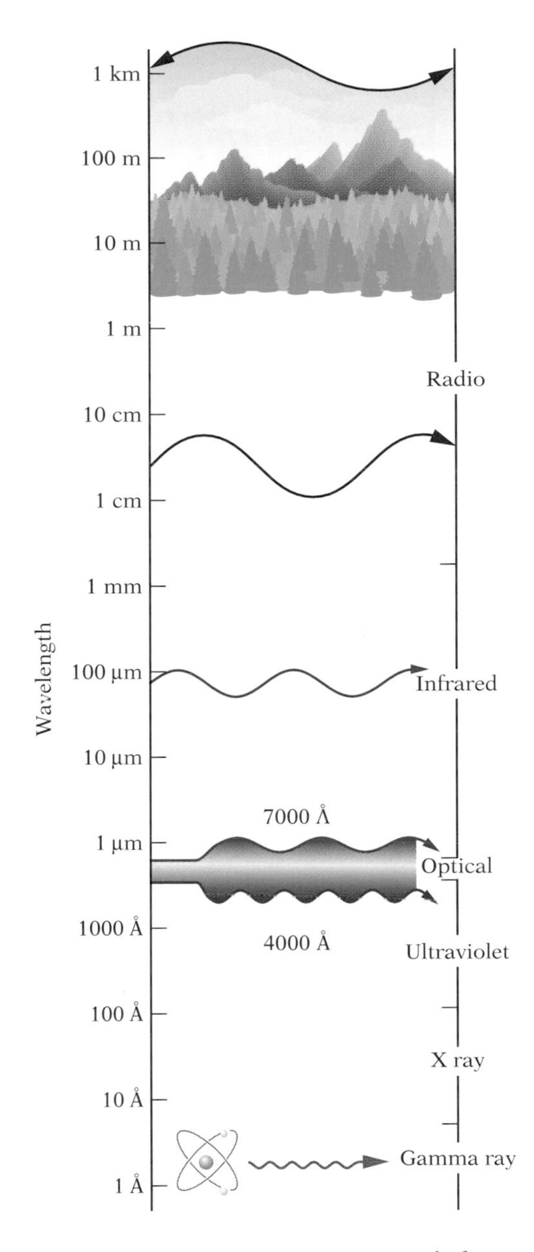

The electromagnetic spectrum extends from gamma rays with the wavelengths of atoms to radio waves that encompass a landscape. Optical (visual) radiation covers only a factor of 2 in wavelength in the middle of the range.

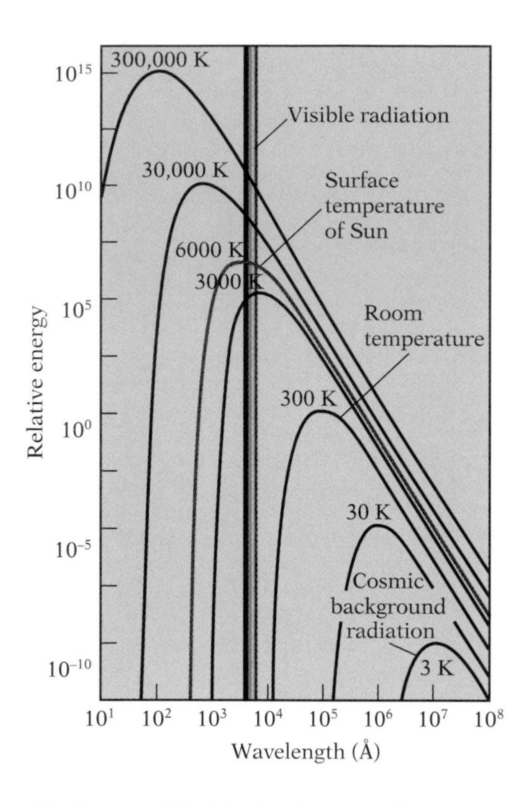

The hotter a blackbody, the more short-wave, high-frequency, high-energy radiation it emits.

only low-temperature radio photons. Even at a miserably cold 10 K, an opaque cloud of interstellar dust radiates strongly at long radio wavelengths. At 500 K, the body can not only emit radio waves but also has the internal energy to generate infrared radiation. At 5000 K, near the surface temperature of the Sun, optical radiation pours out in addition to infrared and radio waves; at 50,000 K, the highest-energy photons are in the ultraviolet; at 500,000 K, in the X ray; and at 5 million K, in the gamma ray.

The effect can be seen merely by turning on a toaster. At room temperature it radiates only in the invisible radio and infrared (as do you—place your hand next to your cheek and feel the infrared radiation pouring from your skin). When the electric current begins to heat the toaster's filaments, you first see them glow a deep red (the color produced by the longest optical wavelengths), then a brighter cherry red, then orange as the metal becomes capable of emitting shorter- and shorter-wave photons. If you could run unlimited power through the toaster, the filament color would move through the yellow, to white (when the maximum is in the green, the combination of color appears white to the eye), and then to bluish, by which time of course the toaster would have melted. Stars behave exactly the same way, except that their energy comes from gravitational compression and the internal conversion of one chemical element into another. The reddish stars must therefore be the coolest (3000 K or so) and the bluish ones the hottest, with temperatures that range to near 50,000 K. White stars fall near 10,000 K and yellow-white ones, like the Sun, near 6000 K.

Newton's simple prism lies at the heart of a device called a spectroscope, which employs lenses to focus the spectrum, allowing the viewer to see an almost infinite, continuously blending array of thousands of shades of color. In 1802, the English scientist William Wollaston discovered dark gaps or lines perpendicular to the flow of colors in the solar spectrum; by 1815, the German optician Josef Fraunhofer had found over 300 such lines. Their significance was discovered some 35 years later in the physics laboratory of the University of Heidelberg, where Robert Wilhelm von Bunsen and Gustav Kirchhoff were working on identifying the elements by the radiation they emitted when heated to incandescence. Using Bunsen's clean new burner, they saw *bright* lines of color in the spectra they produced whose wavelengths uniquely identified the samples.

Fraunhofer had already noted that a pair of bright lines produced by sodium had dark counterparts in the solar spectrum.

Bunsen and Kirchhoff worked systematically to identify a number of other elements whose bright lines could be matched to the wavelengths of the dark solar lines. For example, sodium always produces yellow-orange lines at 5890 and 5896 Å (as well as at many other locations), and magnesium produces lines at 5167, 5173, and 5184 Å. Hydrogen's optical lines fall at 6563, 4861, 4340, and 4101 Å.

The atoms of elemental substances, like those analyzed by Bunsen and Kirchhoff, consist of nuclei made of positively charged protons and neutral neutrons surrounded by (in a very loose sense "orbited" by) negative electrons. Each chemical element has a different atomic number, the number of its nuclear protons: hydrogen has 1, helium 2, carbon 6, uranium 92. Ions are charged atoms from which one or more electrons have been stripped away; molecules are combinations of the same or different atoms. Each gap in the solar spectrum was eventually related to a specific atom, ion, or molecule; 68 of the 90 chemical elements that make the Earth have now been found in the solar gases, as have a variety of simple molecules, including water vapor.

Why the exact reversal of bright and dark lines? Kirchhoff postulated that a substance that emits light at specific wavelengths is also able to absorb it at those same wavelengths. He observed that a hot, dense body—a blackbody—produces a continuous spectrum,

The Sun's spectrum is crossed by thousands of dark absorption lines, each produced by a specific kind of atom, ion, or molecule.

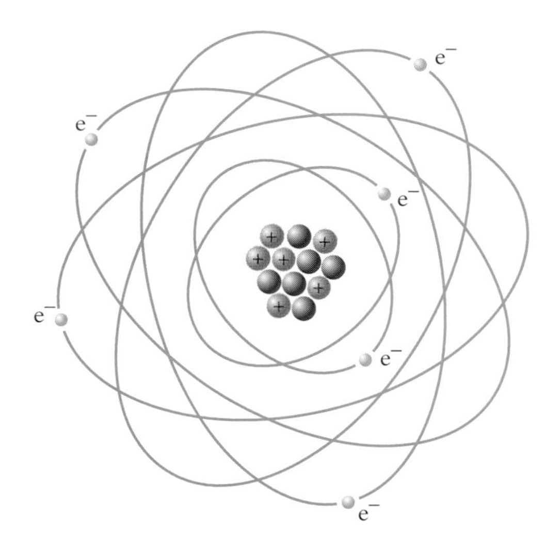

The carbon atom shown has a nucleus with six protons (blue) and six neutrons (black), giving it an atomic weight of 12 and making it ^{12}C. The neutral atom has six negative electrons (green) surrounding the nucleus.

emitting light at all visible wavelengths (and as we know now, also at invisible wavelengths, depending upon temperature). If the source is observed through an intervening substance of lower temperature and density—a cool, transparent gas—radiation is selectively absorbed at wavelengths characteristic of the cool matter, and "absorption lines" are seen against the rainbow continuum. Now the solar spectrum could be explained. Temperature increases inward from the outer solar layers. The continuous spectrum is produced by the hotter, denser, lower layers. As the radiation pours outward, the outer, cooler, more tenuous layers superimpose an absorption-line spectrum. However, if that same gas is not positioned against a brighter background, but instead is observed on its own, "emission lines" at the same wavelengths as the absorption lines are seen. The wavelengths of the lines, whether bright or dark, depend on the chemical composition of the gas, and the relative strengths of the lines—how bright the emission lines, or how dark the absorption lines—depend on temperature, density, and the relative abundances of the different atoms or ions. These observations, a foundation rock of spectral analysis, are known today as "Kirchhoff's laws"; the *mechanism* that produces the lines—absorption or emission—would not be known for another half century.

"A Secret Place of Creation"

Kirchhoff's work made discrimination among the different kinds of "nebulae" possible. Using his principles, astronomers could tell the difference between objects made of unresolved stars and those made of hot, low-density gas. Were the spiral nebulae indeed made of stars? Was the Orion Nebula made of stars or of Herschel's "shining fluid"?

Fraunhofer turned his attention from the Sun to the stars. By the 1820s he had examined a number of the brightest ones and found that they too had absorption-line spectra. Although varying considerably in detail, by their presence the absorption lines at least showed that the stars were constructed along the solar model. Better telescopes and spectroscopes allowed two masters of observational spectroscopy, the English astronomer William Huggins and Angelo Secchi, an Italian Jesuit, to probe much deeper. Though some stellar spectra looked like that of the Sun, others had much more powerful hydrogen lines, while still others had none at all and

Sir William Huggins poses with his telescope and spectrograph; Lady Huggins, née Margaret Murray Lindsay, a talented amateur astronomer since girlhood, actively assisted her husband.

displayed lines of molecules instead. By 1870 Secchi had established the first viable spectral classification scheme for the stars, a forerunner of the system used today. Even though the spectra exhibited remarkable variety, however, they all displayed identifiable lines of the familiar terrestrial elements. Commenting on his own work in 1909, a year before he died at the age of 87, Huggins wrote that "a common chemistry, it was shown, exists throughout the Universe."

Spectroscopic observation of the much fainter extended nebulae was more difficult than was the observation of stellar spectra. Huggins, working in the observatory he had built in his garden at Tulse Hill, London, persevered. The best targets were the brighter

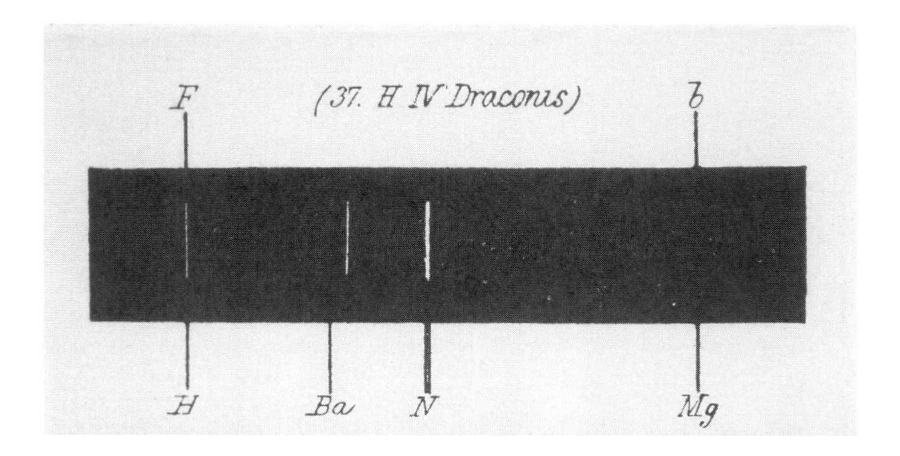

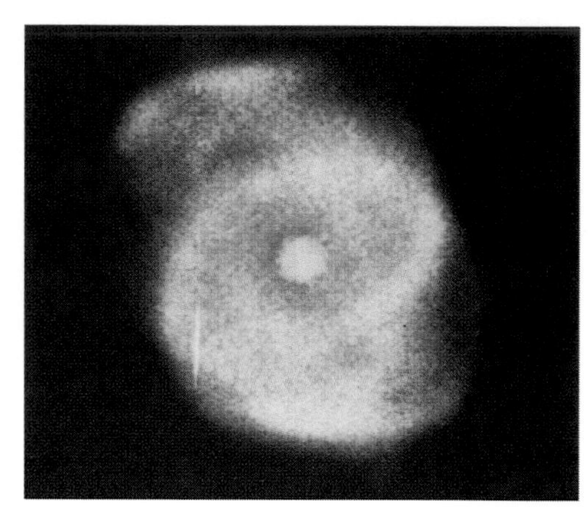

Huggins found that one of Herschel's planetary nebulae, NGC 6543 (at right), radiated three bright emission lines (above), clearly proving that such nebulae are gaseous. He readily identified the left-hand blue line with hydrogen; the other two were not identified until 1928.

planetary nebulae, which at the time were suspected of being stars in formation. Listen to his words, from a memoir written in 1897:

> On the evening of August 29, 1864, I directed the telescope . . . to a planetary nebula in Draco [NGC 6543]. The reader may be able to picture to himself . . . the feeling of excited suspense, mingled with a degree of awe, with which, after a few moments of hesitation, I put my eye to the spectroscope. Was I not about to look into a secret place of creation?
>
> I looked into the spectroscope. No such spectrum as I expected! A single bright line only! . . . The light of the nebula was monochromatic, and so, unlike any other light I had yet subjected to prismatic examination, could not be extended out to form a complete spectrum. . . . A little closer looking showed two other bright lines on the side towards the blue. The riddle of the nebulae was solved. The answer, which had come to us in the light itself, read: Not an aggregation of stars, but a luminous gas.

By 1872 Huggins had also observed the Orion Nebula, in which he found the same three emission lines and another one deeper into the violet. There was no longer any doubt that the Orion Nebula was gaseous too, the brightest of the lines giving the central part of the nebula its greenish cast. Although the longer-wave lines remained unidentified until 1928, Huggins recognized that the other (shorter-wave) pair was radiated by hydrogen; thus the nebular spectrum showed at least some consistency with the compositions of the Sun and stars, a substantive link between the stars and the gaseous nebulae.

Huggins also saw that the Andromeda Nebula (M 31) and its small companion (M 32), unlike Orion, had continuous spectra similar to those of the stars. Even though the absorption lines were not found until 1898, Huggins's observations were consistent with a stellar composition for M 31 and by extension for the other spiral nebulae, the individual stars too faint to be seen. With his spectroscope, Huggins had finally used his observations and Kirchhoff's laws to separate the nebulae into two distinct groups: objects that consisted of stars, even though unresolved, and the gaseous nebulae, which did *not* consist of stars. He thus firmly established the Orion Nebula and the planetary nebulae as true interstellar matter, as the first known components of the interstellar medium, as the first "cosmic clouds."

Edwin Hubble finally resolved the Andromeda Nebula into stars in 1923, and his achievement demonstrated that it, like about a third of the Messier objects and a vast number in Herschel's catalogue, is really an external galaxy lying outside our Galaxy at a great distance (now known to be 730,000 pc, about 2 million ly). This discovery put an end to the sometimes fierce debate about whether the Universe consisted of one "big galaxy"—ours—or whether there were many galaxies similar to ours, "island universes" in the words of Immanuel Kant. Ours was just one of many, of thousands (as astronomers first thought), then millions, now trillions, as far as we can see. The spectroscopic discrimination of the nebulae thus led us inward to begin to understand the nature of our own galaxy, and outward to understand the character of the Universe as a whole.

Like Herschel, Huggins thought that planetary nebulae might be stellar birthplaces; like Herschel, he was wrong. The gaseous nebulae encompass several different kinds of cosmic clouds, each now known to represent a different stage of stellar life, each a specific part of the grand recycling scheme that takes matter from the interstellar medium, deposits it into stars, and releases part of it back into space, leaving behind planets and ultimately dead, burned-out stellar cinders. Decades of research, to be examined in later chapters, have shown that planetary nebulae are not the precursors of the stars, but instead are the precursors of the *deaths* of stars. Stellar youth is associated instead with objects like the Orion Nebula, an example of a *diffuse* nebula. The two are connected through the dark dust clouds that litter the Milky Way, which do not radiate in the optical but in fact impede our view of what takes place inside them. Within these clouds the stars are actually made from ancient matter mixed with the ejecta of evolved stars (which

The opening bars to one of William Herschel's over two dozen symphonies, this one in D major, scored for four violins, viola, violoncello, contrabasso, bassoon, and cembalo.

include the planetary nebulae) and the remains of exploded stars of which Herschel and Huggins were unaware. The diffuse nebulae subsequently mark the sites in which new stars have already been formed, the cycle starting over again. Not until astronomers could peer inside the dark clouds, with infrared and radio instrumentation developed only within the past few decades, could they truly look into Huggins's "secret place of creation."

Into the Darkness

Our view of the Universe and its processes is profoundly influenced by the limitations of the human eye. We see nature in such exquisite and colorful detail that we are easily fooled into thinking that our vision is definitive, that we can be aware of, and understand, all her phenomena merely by looking at them. Such unwitting and natural arrogance is comparable to the geocentric philosophy of Aristotle and Ptolemy that placed the Earth and ourselves at the center of the Universe. Today we understand that not only is the Earth not central to the Solar System, but also that the Sun is only one of billions of stars, set off to the side of a medium-sized galaxy in a poor cluster of galaxies at the edge of a larger cluster. We now also know that our eyes witness only a tiny portion of our world, just one small segment of an observable electromagnetic spectrum that ranges over *15 powers of 10* from wavelengths under an angstrom to greater than a kilometer. Aside from gravitational effects, unless a physical process happens to interact with or produce radiation within the narrow band of 4000 Å to 8000 Å, we will be unaware of it.

The exploration of the electromagnetic spectrum defines the history of twentieth-century astronomy, of the explosion of knowledge in our own time, and is crucial to probing into both the bright and dark depths of interstellar space. William Herschel widened the window to the spectrum in the early 1800s with his discovery of infrared radiation from the Sun, and photography allowed us to probe the small bit of the ultraviolet spectrum that can sneak through the Earth's obscuring atmosphere. But the window did not begin to open fully until 1936 when Karl Jansky, an engineer for Bell Labs looking for a source of communications interference, discovered radio radiation coming from the Milky Way. Improvements in electronics stemming in part from the development of radar in World War II enabled the new field of radio astronomy to expand rapidly in the 1950s. Optical waves are stopped by dust much as they are by terrestrial fog. Radio waves, as well as infrared radiation, however, get through; you have no trouble listening to a radio on a foggy day. Observations in the radio eventually allowed us to determine the detailed conditions in, and the chemistry of, the dark clouds of interstellar space. Previously hampered by the lack of sensitive detectors, in the late 1960s astronomers began to explore Herschel's infrared domain with ground-based and satellite observations that ultimately filled the gap between the radio and the optical. Infrared waves not only penetrate the dust but also are radiated by the dust particles themselves when they are warmed by embedded stars, thus providing a means of probing the nature of the dust and the processes of star formation.

At the other end of the spectrum, orbiting satellites extended our observations to shorter wavelengths that are blocked by the terrestrial atmosphere, through the ultraviolet and into the X-ray domains, where we can witness the results of powerful stellar winds and devastating explosions that help shape the interstellar medium into cosmic clouds and promote the births of stars and planets. Even the gamma-ray spectrum has fallen into our grasp. Astronomers now have the technical ability to go far beyond our human eyes and, in principle, to see the Universe as it really is.

Star Power

The formation of stars from the interstellar medium and the return of matter from stars back into interstellar space is a partially closed cycle that opens up to produce a few unrecyclable objects—notably planets and compact, dense stellar remnants. Buried within the cycle is a powerful engine run by the stars that helps to drive it. That engine exposes more of the intimate symbiotic relationships among stars, interstellar matter, and successive stellar generations.

To see the engine's operation, we start at the surfaces of stars and then dive within them. It intrigued Huggins, Secchi, and the other spectroscopists of the nineteenth century that few stellar spectra look like that of the Sun. The solar spectrum is dominated by powerful absorptions of singly ionized calcium (calcium with one electron removed); the hydrogen absorptions are considerably weaker. The spectrum of the star Vega, however, is dominated by the hydrogen lines, while in the spectrum of Scorpio's Antares the signature of hydrogen can barely be seen. Other stars have no hydrogen lines at all but instead have complex bands produced by

The stellar classification system, shown by nearly 100-year-old Harvard photos, ranges in temperature from hot O stars to cool M. The strong lines in Sirius's A-type spectrum are hydrogen, which weaken both upwards and downwards (they are slightly shifted to the right in ζ Puppis's O-type spectrum, which is on a different scale). Lines of other elements and molecular compounds have different temperature dependencies, allowing a star to be classified quite easily. Those of neutral helium are seen in the B stars and those of both neutral and ionized helium appear in class O. From class F downward we see a steadily increasing number of absorptions produced by ionized and neutral metals (the strong pair to the left are those of ionized calcium). The banded structure to the right in Mira's spectrum is caused by the titanium oxide molecule.

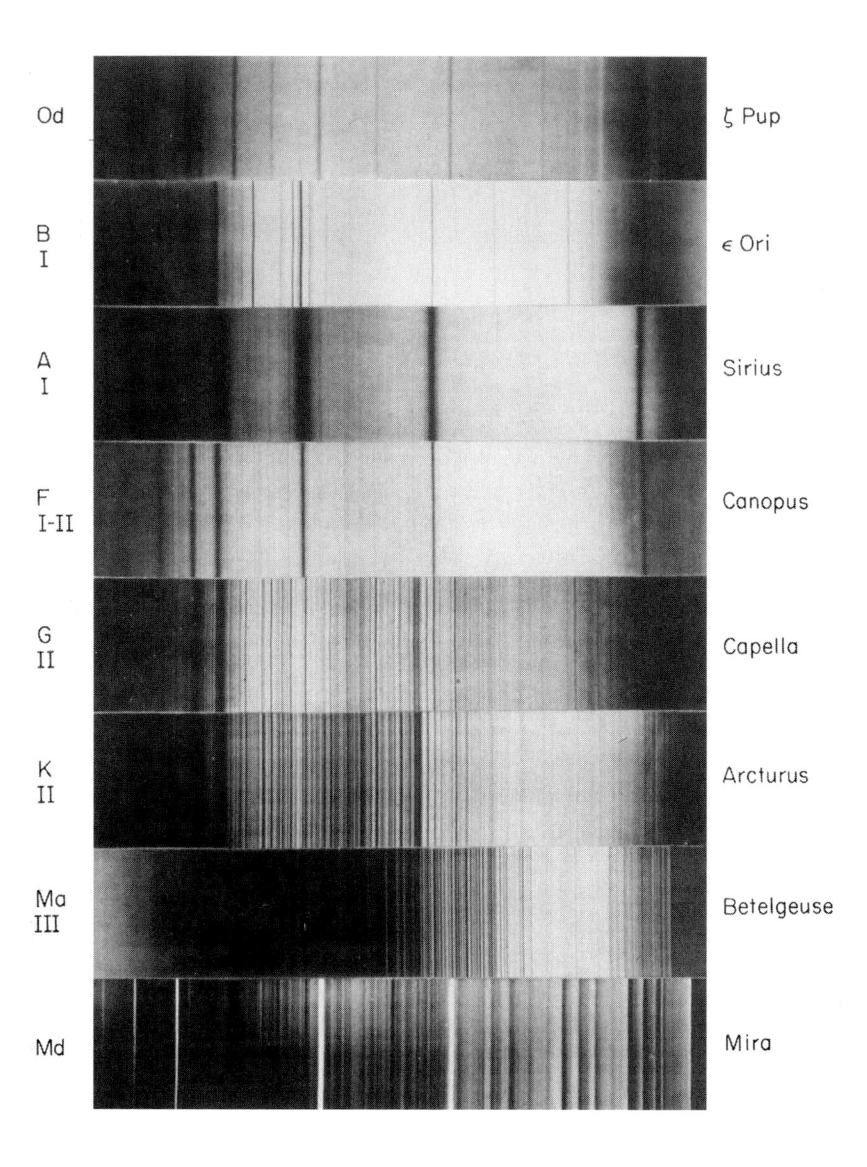

absorptions of a molecule later determined to be titanium oxide, and still others display lines of carbon molecules. About 1900, Harvard's Edward Pickering, working in part from Secchi's researches, ordered the stars in alphabetical classes A through N chiefly by the strengths of their hydrogen lines. By that time, spectroscopes had been largely replaced by spectrographs, their photographic counterparts, which allowed permanent recording of the spectra. As a result of work principally by Pickering's assistant, Annie Cannon, some letter-classes were dropped as redundant or unneeded; in addition, the continuity of other lines from one class to another was best achieved by minor re-ordering. The result was the spectral sequence of seven major classes, OBAFGKM, within which the hydrogen line strengths fall off from class A in both directions. Seven classes were hardly sufficient for the observed gradations, so the classes were decimalized: for example, class A is subdivided A0–A9. The Sun is a G2 star, Vega, Altair, Deneb, and Sirius are all class A, northern spring's Arcturus is K1, and summer's Antares is M1. The classes correlate with color, from bluish at class O, to white at A, through orange at K, and then to reddish at M. Since color relates to surface temperature, so too must the spectral sequence, running from 3000 K at class M to 6000 K at G, to 10,000 K at A, and on to 50,000 K in class O.

From the array of stellar spectra, the first impression is that stars of different classes and surface temperatures are made of different substances. They are not. The physical understanding of the atom in the early twentieth century demonstrated that the differences seen among stellar spectra are almost entirely the result of differing, temperature-dependent, efficiencies of atomic absorption. Absorption and emission lines are caused by changes in the energies of an atom's associated electrons. When an atom absorbs a photon, an electron's energy increases; emission lines are created by photons released when electrons give up energy. Atomic and molecular electron energies are restricted to certain values, hence the related and discrete wavelengths of the spectrum lines.

Hydrogen's electron must be in what is called an excited energy state, already containing considerable energy, before it can absorb lines at optical wavelengths, the "optical lines"—in other words, the atom must be in a warm gas before the lines can be seen. The atmospheres of the coolest stars therefore display no hydrogen lines. As we move through the spectral sequence from M to K, through G and F, to A, the stars become hotter and the hydrogen lines become more prominent. At temperatures hotter than class A, in the B and O stars, atomic collisions are so violent that the hydrogen atoms become ionized, their lone electrons removed, and the hydrogen line strengths are again reduced. All atoms and ions have such temperature dependencies. When they are taken into account, we find that almost all stars have the same chemical compositions, supporting Huggins's brave contention that a common chemistry reigns throughout the universe. We find that, in the vast majority of stars, for every 1000 hydrogen atoms there are 80 of helium and only 1 that represents the proportion of all the heavier elements. In general, the heavier the element the less there is of it, though there are

In general, the heavier the atom the lower its abundance in the Universe, though there are sharp peaks at iron and lead and a sharp dip at lithium, beryllium, and boron.

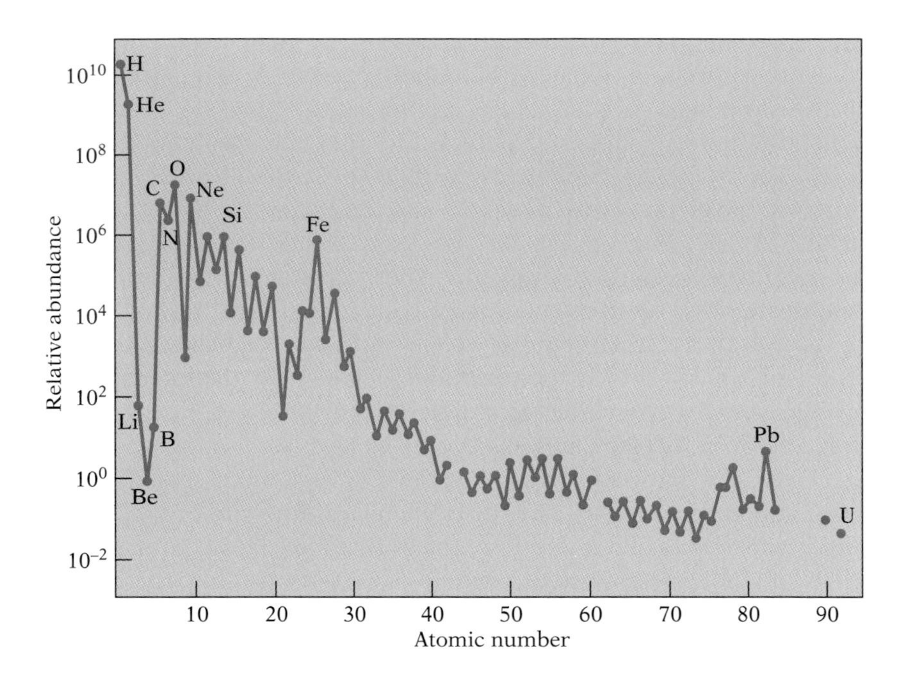

some interesting exceptions (for example, lithium, beryllium, and boron are very rare, while iron and its related metals are relatively abundant). Except for hydrogen and helium, the mix is closely the same as that found on the Earth, demonstrating a unity, through the stars, between the planets and the interstellar medium.

Knowledge of the high abundance of hydrogen in stars was the key to understanding solar and stellar power sources. The Sun, a gas throughout, is held together by its own gravity, its interior heated by gravitational compression. The British astronomer Arthur Eddington knew that the nucleus of a helium atom is slightly lighter than the combined nuclei of four hydrogen atoms, each one of which is a single proton. He pondered that if four hydrogen nuclei could be fused in the high temperature of the solar interior into one of helium, the lost mass would be enough, through Einstein's $E = Mc^2$ (c is the speed of light), to power the Sun. Since astronomers, particularly Harvard's Cecilia Payne, had demonstrated that the stars were mostly hydrogen, there was an abundant fuel supply. Discoveries in atomic physics that included the neutron, the positron (a positive electron with reversed electric charge), and the neutrino (a massless or near-massless particle that carries energy) ultimately enabled the physicists Hans Bethe and Charles Critchfield to describe the proton–proton chain in which four atoms of hydrogen (that is, four protons) are combined in succession into one of helium. Since the helium nucleus consists of two protons and two neutrons, the hydrogen nuclei must get rid of their positive charge, which they do by spitting out positrons (as well as neutrinos). The positrons mutually annihilate electrons with the creation of gamma rays, the ultimate result the optical radiation that pours from the Sun.

Even though protons have positive charges and naturally repel one another, they (and the neutrons) are held together by a powerful short-range nuclear force, which within a distance of 10^{-13} cm overpowers electrical repulsion. Fusion of hydrogen into helium can take place only where the temperature is sufficiently high, above about 10 million K, to ram nuclei together close enough for the strong force to lock the particles together; such high temperatures exist in a solar core that encompasses about 40 percent of the Sun's mass. We calculate that the Sun began its life with enough hydrogen to keep it "burning" (in the nuclear sense) at its present level for a total of about 10 billion years. From the dating of rocks from the Earth, Moon, and interplanetary space, we know that the Sun is already 5 billion years old and therefore has another 5 billion to go, after which it will begin to die. When it does it will produce a planetary nebula that will quickly dissipate into interstellar space.

As stars age, the internal temperatures can rise so high that the fusion engine can manufacture heavier elements. Stars like the Sun create atoms up to carbon and oxygen. The most potent furnaces in the celestial chemical factory, however, are the O stars, several of which are found in double star, or binary, systems. Application of Newton's generalization of Kepler's third law to the orbital characteristics allows us to determine stellar masses. The O stars, though quite rare, are the most massive of all stars, at maximum reaching a

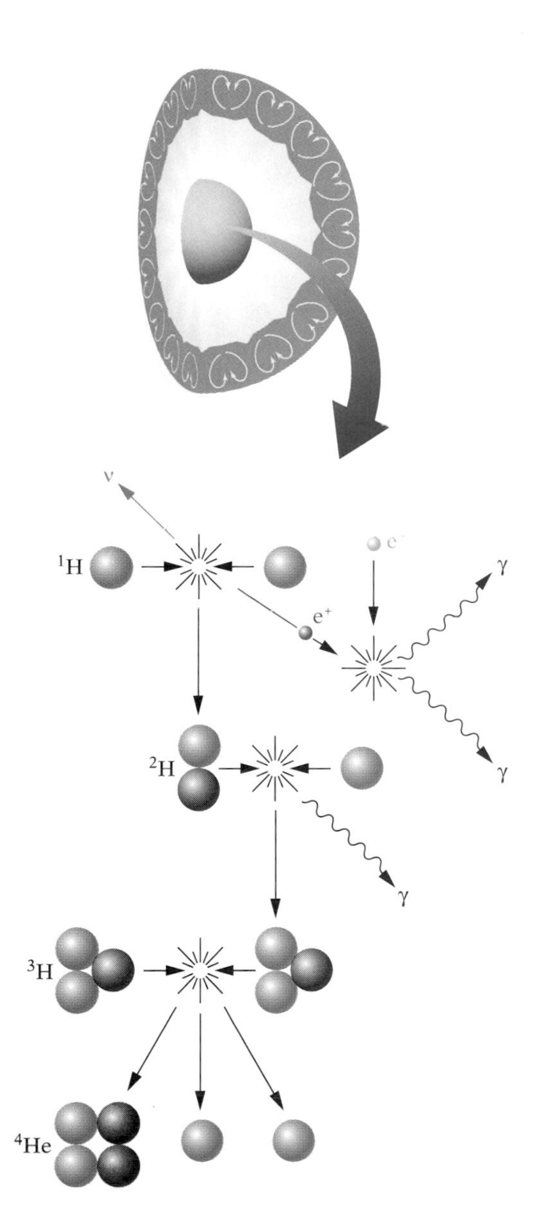

The Sun's energy is generated principally in its hot (15-million-K) core by the proton– proton chain. Pairs of hydrogen nuclei—protons—fuse, each producing a deuterium nucleus (^{2}H, hydrogen with a neutron attached) as one proton converts to a neutron with the ejection of a positive electron, a positron. The positron is annihilated by a negative electron, the event producing two gamma rays. A neutrino (ν) is also released in the reaction. The deuterium then captures another proton to make another gamma ray and a light helium nucleus (^{3}He, a helium atom with a single neutron); two ^{3}He atoms fuse into ordinary helium, ^{4}He. In transmitting the radiation, the outer solar layers degrade it from gamma rays into optical light and serve to insulate the hot interior.

hundred times the mass of the Sun. Their great internal heat generates an enormous outpouring of energy, making them the most luminous of stars, some shining at a million times the solar luminosity. They can get so hot inside that they can fuse nuclei all the way to iron.

We know from observations of planetary nebulae that stars like the Sun lose mass back into space and that some of that mass is enriched in the by-products of nuclear burning. The O stars, however, are destined to explode, and in their fierce detonation they will launch into the cosmos vast numbers of heavy nuclei all the way through the list of elements to uranium and beyond. As a result, the interstellar medium is continuously being enriched in heavy stuff, the variety of fusion reactions creating essentially all the chemical elements heavier than helium. The debris of stellar evolution subsequently mixes with the gas and dust already there. The fusion engine makes the raw materials for the ubiquitous interstellar dust, in whose dense nests the stars are born. Thus the Earth and planets are the distillates not just of the interstellar medium, but of earlier generations of stars.

The O and hottest B stars are communal, clumped together in great associations that are commonly allied with diffuse nebulae. Most of Orion, with its hot blue-white stars and its great nebula, is such a gathering, as is his mythical enemy Scorpius. Clusters are held together by gravity. The "OB associations," however, are not. Though stars are commonly referred to as "fixed," in reality we can watch them move relative to each other through tiny angles measured in fractions of a second of arc per year, the movements in time—in millions of years—destroying the familiar constellations. Over the years we can watch the stars of an OB association move apart, away from a common center, as the association expands in angular dimension. Knowledge of the distances to the associations allows the conversion of angular speed across the line of sight to actual velocity in km/s. From the maximum sizes of the OB associations, we see that the O stars never get far away from their centers of origin before they die, and they must therefore be very young, at most only a few millions of years old.

Diffuse nebulae are almost always visually associated with hot O stars—the best example the O and B star Trapezium inside the Orion Nebula—providing powerful evidence for the relation between them and star birth. This relationship also shows that the stars must somehow be responsible for the illumination of the nebulae. The evolution and eventual explosions of the O stars wreak havoc on their birth clouds, ultimately ripping them to shreds, and in turn the O stars provide interstellar blast waves that will help compress yet other clouds, leading to yet more stars. The stars, and ourselves, therefore cannot be separated from the interstellar clouds nor from other stars as well.

To find how this remarkable integrated system works, we first pick apart the most obvious of the cosmic clouds, the diffuse nebulae, then plunge ever deeper into the dark dusty clouds where we will eventually see an intricate molecular chemistry and can watch the stars and planets being born. The flow of stellar evolution then takes us to the emerging clouds of the planetary nebulae, to the powerful remnants of exploded stars, and back once again into the cosmos, where we return to witness new stellar generations, these the progeny of the past.

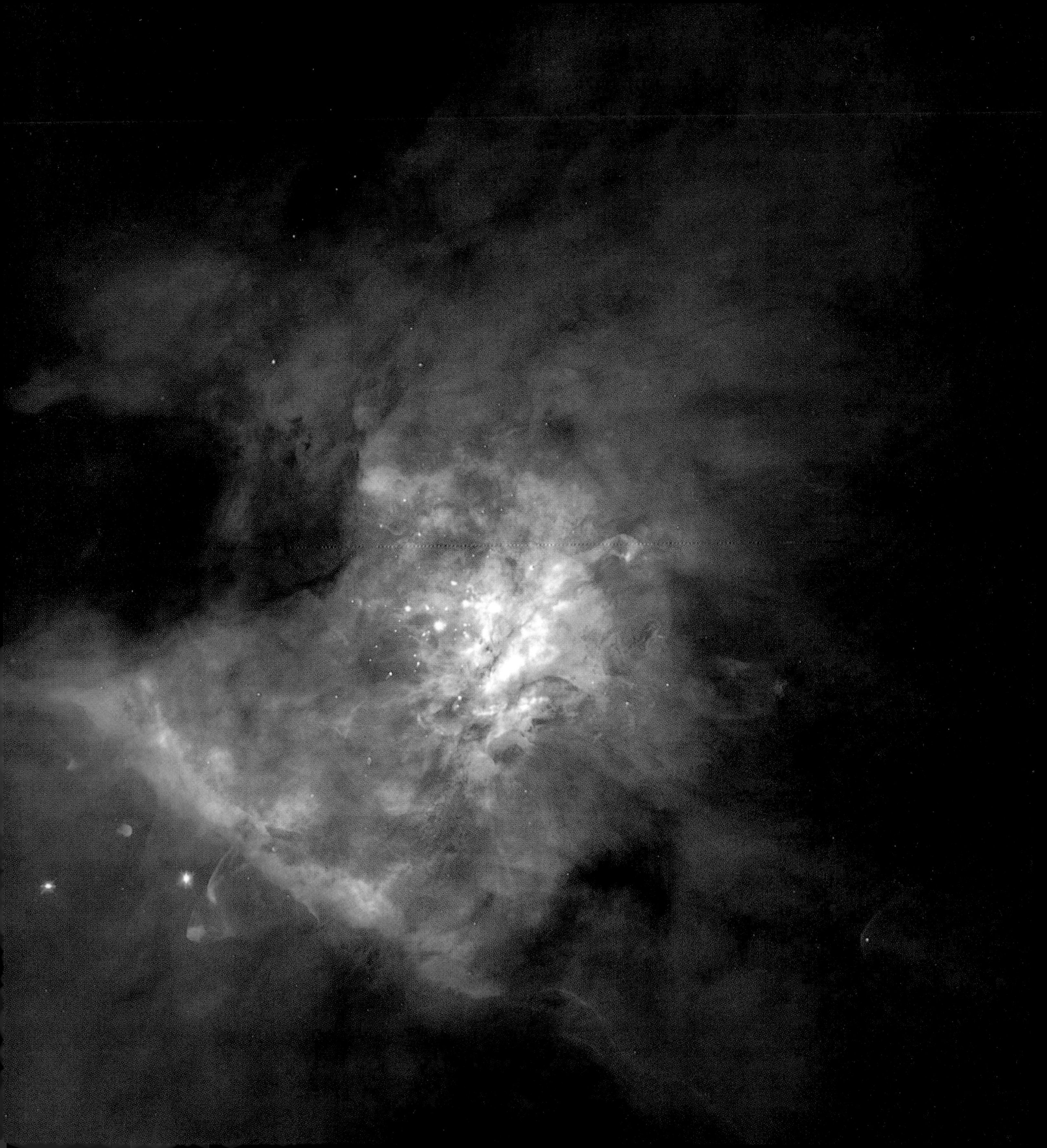

2

Entering the Nebulae

≺ Would that G. P. Bond could see this magnificent composite image of the central Orion Nebula taken by the Hubble Space Telescope. The colors, adjusted to match what would be seen by the eye, are produced by emission lines at different wavelengths. The Trapezium, at center, almost completely dominated by θ^1 Orionis C, has ionized the gas. The "bright bar," obvious at left, is an ionization front that is eating its way into a neutral cloud. Numerous faint stars are seen within, the nebula marking the place of an extraordinarily rich breeding ground.

The most immediately recognizable parts of the interstellar medium are the diffuse nebulae, whose luminous glows render them among the loveliest of astronomical sights. Because of their bright and obvious emission lines, and an atomic theory that could explain them, the collection of diffuse nebulae was also the first component of the interstellar medium to be understood. Conversely, observation of nebular spectra allowed scientists to test various aspects of atomic theory.

We therefore look at the nebulae and the atom together. In doing so we will examine atomic processes that will lead us into a great variety of cosmic clouds. At the same time, we will construct a picture of nebular processes that will be important throughout the cycle that takes us from the dark clouds, to star birth, through stellar ejecta, and back into the dark clouds again.

Markers in the Medium

The diffuse nebulae, epitomized by the Orion Nebula, are among the finest telescopic and photographic showpieces of the sky. Of greater significance, the shining prominence of the diffuse nebulae reveals the galactic distribution of the interstellar medium by providing tracers for the clouds that spawn the stars: since the diffuse nebulae are associated with hot young stars, in fact are lit by the young stars that were created within their gases, they locate regions in the Galaxy where star formation has most recently taken place.

Turn from northern winter and the Orion Nebula to the sights of summer, and examine the Milky Way as it plunges ever brighter toward the southern horizon. There, in Sagittarius, we see a much larger gaseous cloud, the Lagoon Nebula, M 8. Visible to the naked eye, with large binoculars it presents a fine sight. A telescope shows a complex elongated system with a dark bay—the "lagoon"—running through its middle. Inside the nebula, analogous to the Trapezium of the Orion Nebula, is an open cluster of stars that, because it contains young stars, must recently have formed out of the matter in which the stars are now embedded. Sagittarius and its neighboring constellations are littered with such sights, many of them, like the Trifid, the Omega, and the Eagle, named for their appearance.

Countless fainter nebulae display enormous variety. Northern summer prominently features the North America Nebula (NGC 7000) in the constellation Cygnus near the star Deneb. Its size (over a degree) and low surface brightness render it a difficult object for the telescope. But in a very dark sky it is rather obvious in binoculars and can actually be seen with the naked eye once you know it is there. The far southern hemisphere is equally rich in diffuse treasures. To the west of the Southern Cross, in the sprawling constellation Carina (the "keel" of the great ship Argo), lies the magnificent Carina Nebula, which is more than 2° wide and readily visible to the naked eye. Within its complex confines are a vast number of bright O and B stars (and far more lesser stars) that include one of the Galaxy's most luminous and massive stars, Eta Carinae.

A singular characteristic of our Galaxy's bright diffuse nebulae is that they lie strictly within the Milky Way (providing it with much of its beauty and charm), and thus they are within the plane of the Galaxy, the galactic disk. Fainter diffuse nebulae are *everywhere* in the Milky Way, huge sheets of weakly glowing diffuse gas that have no clear boundaries. Set within this background are the famous named objects, vast numbers of others, and innumerable tiny

Sagittarius is rich in diffuse nebulae, the greatest of them the Lagoon Nebula.

Northern Cygnus features the North America Nebula, with its obvious east and west coasts and Gulf of Mexico. The Pelican Nebula lies to the right, faint background nebulae suffusing the rest of the picture.

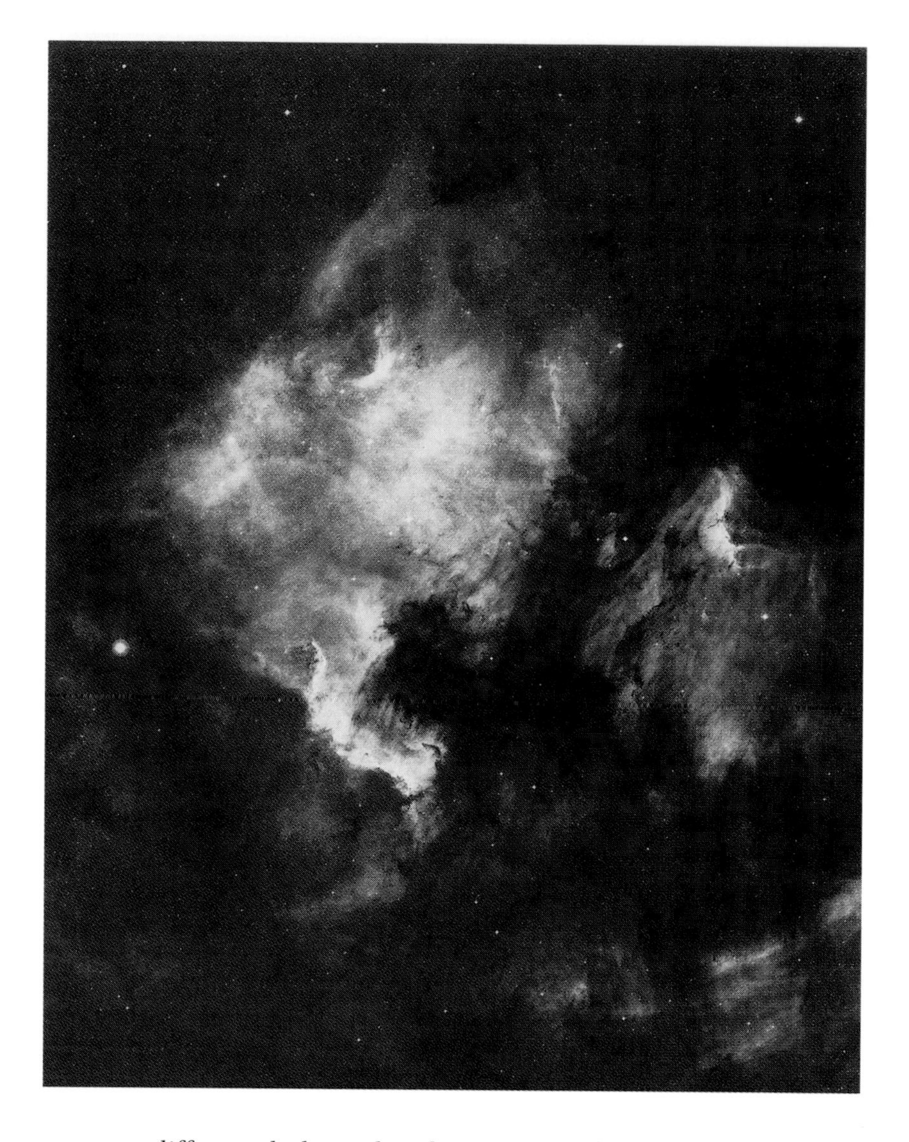

compact diffuse nebulae only a few minutes of arc across. Whatever form they take, the diffuse nebulae are invariably located in and around dark, obscuring clouds, the juxtaposition clearly linking the nebulae with the dusty component of the interstellar medium—and again with star formation.

Diffuse nebulae throng other galaxies too. More magnificent than anything in the Milky Way is the Tarantula Nebula in the Large Magellanic Cloud, a small, nearby galaxy that is a companion to our own. The Tarantula is so large that even though 52,000 pc away it is half a degree across. Within it lie hosts of O-star clusters. The great diffuse nebula NGC 604 in the much more distant galaxy M 33 is comparable. How ironic that the nineteenth century's "spiral nebulae"—like M 33—which had been once been confused with the diffuse nebulae, should themselves *contain* diffuse nebulae. Looking at other disk-shaped galaxies, those structured like our own, we see not only that the diffuse nebulae are confined to the disks but also specifically to their spiral arms. The arms are caused by giant density waves that move through the disk, compressing its matter. They are related to a galaxy's rotation, the waves starting near the galaxy's center and opening outward as the disk rotates. Since the diffuse nebulae mark the location of interstellar matter, the bulk of it appears to be contained by the spiral arms. Because our Galaxy has a disk, it too ought to have spiral arms, though they are not immediately evident to us because we are inside our Galaxy; the arms must surround us on all sides, and we can have no broad perspective. However, by analogy, they should be outlined by our own diffuse nebulae. Because O stars are young and commonly associated with diffuse nebulae, they too should mark the locations of the arms. Maps of diffuse nebulae or O stars will therefore tell us where the local arms are. Moreover, since the O stars are young, and these and the diffuse nebulae are associated with spiral arms, the arms must have something to do with star formation.

Though in another galaxy 700,000 pc away, the giant diffuse nebula NGC 604, 500 pc across, is easily visible from Earth. The Hubble Space Telescope reveals it to be an enormously complex object, full of filaments and loops, lit by the light of 200 or more massive stars.

However, argument by proximity, while important and suggestive, does not establish true physical relationships. For that we need dig deeper to understand why and how the young stars can excite the nebulae to glow, how the nebulae and their associated interstellar clouds can produce stars, and ultimately to appreciate the specific role played by the arms in star formation. To begin to establish these concepts, we look at what the embedded stars can tell us about the nebulae, and then at the intimate relations among the stars, the nebulae, and the atom.

Distance and Dimension: A Gift from the Stars

We cannot establish the natures of the nebulae without knowing how far away they are; their distances give us their dimensions and place them into their galactic contexts. All we need do to get the distances to the nebulae is to find those of their embedded stars. Unfortunately, all such stars are so far away that their parallaxes are too small to measure. We need another way to determine distance, which is found in the characteristics of the stars themselves and within the spectral sequence, OBAFGKM, that classifies them.

The story, in this most ancient of sciences, began over 2000 years ago with the first attempts to measure (or rather, to estimate) stellar properties. In 150 B.C., the Greek astronomer Hipparchus grouped the sky's stars into six brightness categories called magnitudes, first magnitude representing the brightest stars he saw, sixth the faintest. His system survives, though in a quantitative and logarithmic form in which first magnitude is set to be 100 times brighter than sixth. If five magnitude divisions correspond to a factor of 100 in brightness, a single whole magnitude division corresponds to a factor of the fifth root of 100, or 2.512. . . . Each magnitude division is therefore 2.512 . . . times brighter than the next fainter division. The human eye can easily discriminate a tenth of a division, and modern detectors nearly a thousandth. With the faintest stars averaging 6.0, the magnitudes of the very brightest stars, like northern winter's Sirius, must be expressed by negative numbers. With optical aid we see stars fainter than sixth: binoculars show eighth, and the best telescopes extend our range nearly to the thirtieth. The small range of numbers in the magnitude system is deceptive. Since each set of 5 magnitudes corresponds to a factor of 100 in apparent brightness, a difference of 10 magnitudes (two sets of 5) corresponds to a factor of 100 × 100, or 10,000. The faintest stars that can be observed by telescope are thus over a billion times dimmer than those visible with the unaided eye. The stars of Orion's Trapezium range between fifth and eighth.

The *apparent* magnitude (m) of a star—the magnitude as seen by Hipparchus and by us—depends on the star's real luminosity (that is, on the amount of energy it radiates in watts) and on its distance. A star may seem bright to us either because it is nearby (even though faint) or because, though far away, it is very luminous. If we know the star's distance, we can find its luminosity from its apparent brightness. In astronomy, luminosity is expressed by a star's *absolute* magnitude (M), the value the apparent magnitude would have at a standard distance of 10 pc or 32.6 ly. The two kinds of magnitude are related by the magnitude equation, $M = m + 5 - 5 \log d$, where d is in parsecs.

With distances known from parallaxes or some other means and m known from electronic measurement at the telescope, we can calculate the absolute magnitudes of vast numbers of stars, including the Sun. Though its luminosity is 4×10^{26} watts, place it 10 pc away and its magnitude would shrink to 4.83, and it would appear about as bright as the faintest star in the Little Dipper, close to the edge of human vision. The Sun's luminosity and absolute magnitude allow the calibration of the entire absolute magnitude scale in terms of stellar luminosity in watts. The range of stellar luminosities, as found through absolute magnitudes, is astonishing. The brightest stars have absolute magnitudes of –10, 15 magnitudes (or a million times) brighter than the Sun. Such a star 10 pc away would be readily visible in daylight and cast obvious shadows at night. At the other end of the scale, the faintest stars are 15 absolute magnitudes dimmer than the Sun, so faint that they would have to be near the outer edge of the Solar System just to be visible to the eye. And these magnitudes—called visual magnitudes because they are based on the physiological response of the human eye—are blind to radiation that falls outside the narrow optical domain of the spectrum. When we take all stellar energy into account, the brightest stars approach 10 million solar luminosities, and for brief explosive periods can far exceed that.

We now have two readily observed properties of a very large number of stars: their luminosities, measured through their distances, and their temperatures, as represented by spectral classes. The combination of the two opens wide the door to the Galaxy. Theory, confirmed by measurement in the physical laboratory, shows that the energy radiated by a blackbody per unit area (for example, in watts per square meter, w/m^2) depends on the fourth power of the temperature. At 5780 K, the Sun radiates a power of 6340 w/m^2; double its temperature to 11,600, into the realm of the B stars, and the radiant energy per m^2 would climb by a factor of 16. It is therefore only natural to graph stellar luminosity against temperature to see what happens. The first such correlations were made, independently, in the early part of the twentieth century by Ejnar Hertzsprung and Henry Norris Russell, and the modern plot is named the Hertzsprung–Russell, or HR, diagram. In its various forms it is the single most important tool in stellar (and by extension, in galactic if not even extragalactic) astronomy: as well as allowing us to extend our measures of distance throughout the Universe, it provides a solid observational platform on which to set theories of stellar evolution, the aging processes of stars.

As usually graphed, the majority of the stars lie on the diagram in a long diagonal, the "main sequence," from lower right (low temperatures and low luminosities) to upper left (high temperatures and high luminosities); in accord with the laws that govern radiation, stellar luminosity increases sharply with increasing temperature. But there is a wonderful surprise, one that ultimately showed astronomers the way stars evolve and eventually die: *another* heavily populated band on the diagram that begins near the center of the main sequence and goes up and to the *right* toward lower temperature. These stars are luminous *even though they are cool.* Two stars with the same temperature release the same amount of radiation

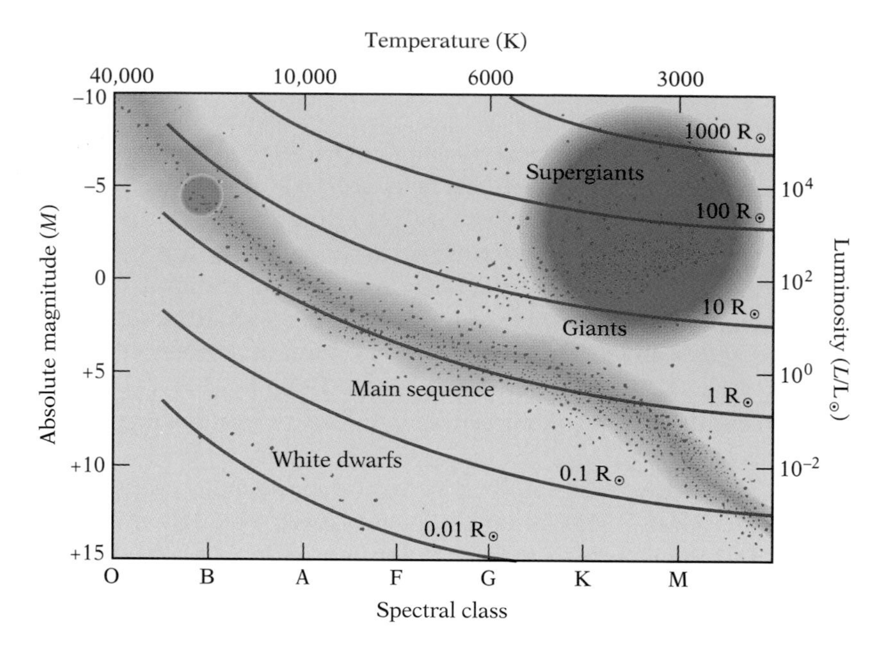

The Hertzsprung–Russell diagram plots stellar absolute magnitudes against spectral classes. Stars strongly concentrate to a main sequence in which mass increases from lower right to upper left. The giants must be large to produce high luminosity at low temperature; the supergiants across the top must be even larger. Analogously, the white dwarfs have to be very small to produce their low luminosities. Lines of constant radius are drawn on the diagram; relative sizes are shown by the colored spheres. The symbols $L_{\odot}$ and $R_{\odot}$ represent, respectively, the solar luminosity and radius.

per unit area. If one of them is brighter than the other, it must have a larger surface area (and radius). Hertzsprung originally distinguished the two groups by calling those on the main sequence "dwarfs" and the larger ones "giants" (oddly, he considered neither type "normal"). To be so bright, the giants must typically be tens of times the size of the Sun; some would reach from the Sun (which is 150 million km, or 0.01 AU, across) past the Earth, even past the orbit of Mars

Sprinkled across the top of the HR diagram are a handful of yet brighter cool stars—the "supergiants"—that must be even larger than the giants. The greatest of these monsters would nearly fill the orbit of Saturn, 20 AU across. Finally, down in the lower left-hand corner is a string of stars, some of them quite hot, that radiate feebly compared even with the dwarfs. In the same line of reasoning, these "white dwarfs" must be very small, about the dimension of Earth. The largest stars have radii 100,000 times the radii of the smallest.

If we discover that one green tree uses chlorophyll to provide it with sustenance, we may assume that all green trees operate the same way. Because we have evidence (from the age of the Earth, the composition of the Sun, and fundamental physics) that the Sun is a natural fusion engine, so too must be other stars; and since the Sun is part of the main sequence, the main sequence ought to be a domain of hydrogen burning. But if main sequence stars all operate on the same principle, what causes their enormous range in luminosity? Binary stars provide the answer. Binary components orbit each other under the influence of their mutual gravity. Their gravity, and hence their orbits, are dependent on stellar mass. From Kepler's laws of planetary motion, as generalized by Newton to all two-body orbits, astronomers have derived hundreds of stellar masses. The masses of stars range from a minimum of about 0.08 solar masses, or 8 percent that of the Sun (which itself is 333,000 times more massive than the Earth), up to the O-star level of around a hundred Suns. The main sequence is physically a *mass* sequence; along it, the greater the mass, the brighter the star. Luminosity roughly depends on the mass to the 3.5 power ($L \propto M^{3.5}$): a dwarf star with twice the mass of the Sun is about 10 times as bright. Greater mass means greater gravitational compression, higher internal temperatures, more energy that can be released as heat and light, and faster rates of nuclear fusion. The youth of the massive O stars (which never get far from their birthplaces before they die) demonstrates that more massive stars live for shorter periods of time, because of the rapid pace of nuclear burning. Less massive stars, however, like the Sun and those farther down the main sequence, can expect to enjoy long lives. Thus they have the time to move away from their birthplaces and any stellar siblings that may have been created with

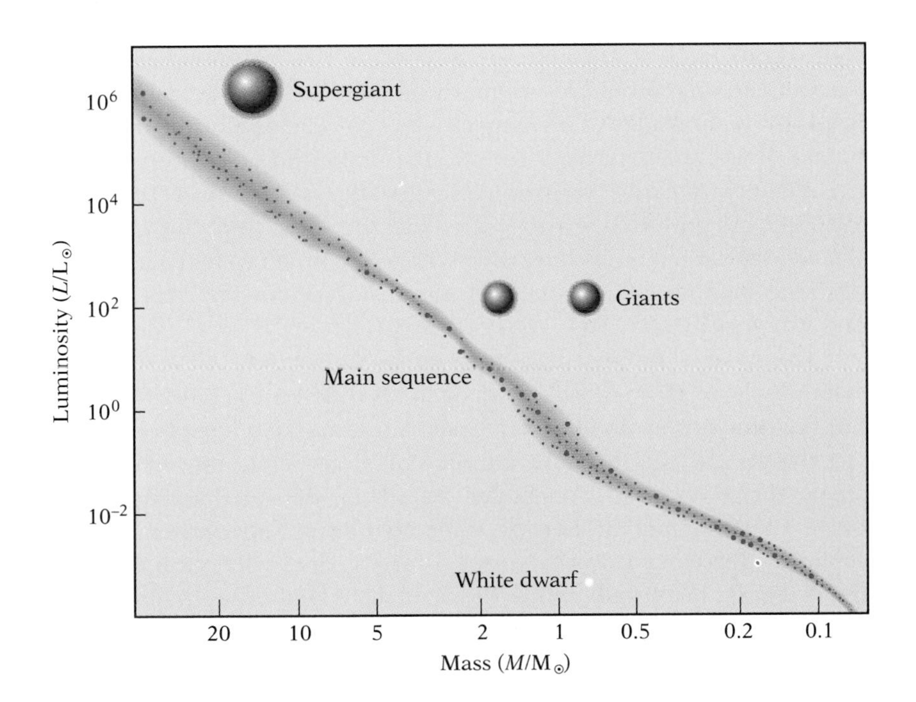

Most stars follow a close relation in which luminosity climbs quickly with mass. Dying stars, however (giants, supergiants, and white dwarfs), fall distinctly off the relation. The symbol M☉ represents the solar mass.

them, and can orbit independently around the galactic center. As a result, most low-mass stars have little direct relation with the diffuse nebulae, which mark active or recent sites of star formation.

Giants and supergiants are brighter than their masses would predict and white dwarfs are enormously fainter, so for these stars something other than simple hydrogen burning must be involved. We now know that, as the main sequence is descended from the interstellar medium, so the giants and supergiants are descended from the main sequence (and so the white dwarfs from the giants). Giants and supergiants eventually make the planetary nebulae and other cosmic clouds that send matter back into the space to feed yet more clouds, out of which will develop stars and more diffuse nebulae, the cycle continually repeating itself.

From their parallaxes and apparent magnitudes, astronomers have learned the luminosities—expressed by absolute magnitudes—of the different spectral classes that populate the middle and lower parts of the main sequence, as well as those of many giants. The luminous O stars and supergiants, however, are quite rare, and none happens to be close enough to allow us to measure its parallax. How then can their luminosities be found? We are saved by the open clusters. We know from their mutual gravitational bondage that the stars of an open cluster must have been born at the same time. Several young clusters, some associated with diffuse nebulae, contain not only entire main sequences of stars (that run from the bottom of the HR diagram right through the O stars) but supergiants as well. From local studies we already know the luminosities of the stars of the middle and lower main sequences. If we determine the apparent magnitudes of a young cluster's middle and lower main sequence stars, we need only reverse the magnitude equation to find the cluster's distance. In turn, this distance allows us to find the absolute magnitudes of the cluster's O stars and supergiants. Decades of such studies have allowed astronomers to learn accurate absolute magnitudes of the stars in all the various spectral classes and categories.

The spectra of the stars in turn tell us their kinds. All A stars, for example, have strong, dark, hydrogen absorption lines. Darkness is not the only criterion, however. Spectrum lines also have breadth in wavelength, in part because atomic collisions jiggle the orbits and smear their energies. The smaller A stars are denser than the larger ones. Because higher density leads to closer approaches among atoms and greater orbit-perturbing forces, the absorption lines of the A dwarfs are broad, while those of the A giants are narrow and of the A supergiants narrower yet. Other spectral classes have similar criteria that give away their luminosities and hence their

The giant Tarantula Nebula occupies a prominent place in our Galaxy's biggest companion, the Large Magellanic Cloud.

absolute magnitudes as read from the HR diagram. Once again, given the absolute magnitudes and measured apparent magnitudes, we can find the distances, now not of clusters, but of any single star we wish. From this powerful method of "spectroscopic distances," θ^1 Orionis C, the luminary of the Orion Nebula's Trapezium, is found to be 450 pc (1300 ly) distant. The Orion Nebula's angular diameter of about 1° and simple trigonometry then yield a physical diameter of 8 pc. To place this glowing gaseous cloud in some kind of perspective, a ray of light takes over 25 years to travel from one side to the other; the nebula could contain the distance between our Sun and the nearest star, α Centauri, six times over.

The Galaxy's other classic, named, diffuse nebulae are roughly the same size. The Tarantula and NGC 604, however, are stunningly different. If the Tarantula were as near to us as the Orion Nebula, it would appear to us to fill all of Orion, and since it is some 300 pc across it would reach a good ways to the Sun. At the other end of the scale are the small, compact, diffuse nebulae that measure only a fraction of a parsec across and are no larger than the bigger planetary nebulae.

The measurement process beautifully demonstrates the remarkable synergy of astronomy, and of science in general. At one time, not all that long ago, star clusters were commonly confused with diffuse nebulae (indeed, also with galaxies). Now, the clusters provide a means by which we can begin to understand not just the diffuse nebulae, the leavings of star formation, but ultimately the cosmic clouds that produced the stars and clusters in the first place.

Atoms and Nebulae

What physical processes allow the O stars to light the diffuse nebulae? We cannot understand the nebulae until we know. The key to the illumination mechanism of both the diffuse and planetary nebulae lay in their emission-line spectra—the radiation of light by specific atoms and ions at specific wavelengths—and in the growing understanding of how the atom works. The knowledge acquired would later be applied to other aspects of the interstellar medium, to other cosmic clouds, allowing astronomers eventually to see how the whole interstellar structure is assembled.

The first step is chemical identification, so we can see just what it is that we have to explain. The three bright emission lines initially discovered by William Huggins in the spectrum of the planetary nebula NGC 6543 (and in the spectrum of the Orion Nebula) have wavelengths of 5007 Å, 4959 Å, and 4861 Å. The other line Huggins had found in Orion lay at 4340 Å. This one and the 4861 Å line (now referred to as Hβ and Hγ respectively) were quickly matched with two lines in the spectrum of hydrogen observed in the laboratory. The other two emissions presented a puzzle. Huggins initially identified the bright line in the green (that at 5007 Å) with nitrogen, even though the laboratory line is double, whereas that from the nebulae is distinctly single, and no other nitrogen emissions were present. He had no suggestions at all for the middle line at 4959 Å. Consummate astronomer that he was, Huggins was the first to photograph a nebular spectrum, and in an 1882 study of the Orion Nebula he found yet another mysterious line in the invisible ultraviolet. Improved observations of both planetary and diffuse nebulae eventually turned up more hydrogen lines, including one at 6563 Å, now called Hα. The emission lines initially told Huggins that the nebulae were made of gas. If astronomers could learn how the lines were produced, they could possibly determine conditions in the gas and begin to understand how the nebulae were formed.

In 1913, the Danish physicist Niels Bohr introduced a theory of the atom that would eventually allow such deductions. The electrons of an atom or molecule are placed in any of an infinite series of allowed "orbits" of different radii. Electrons can exist only in one of these allowed orbits, never between them. The orbital structure of any given kind of atom or ion is the same anywhere in the Universe. However, the electronic and orbital structures of *different* kinds of atoms and their ions are all different. Each ion of each atom is unique. Neutral hydrogen is simple, as there is only one electron to consider. In atoms heavier than helium (whose neutral state has two electrons), the increasingly numerous electrons are arranged in a series of "shells." The emission spectra considered here deal only with the outer electrons (in chemistry these are called the valence electrons)—that is, we ignore the inner ones.

If you lift a bowling ball, you give it potential energy, that is, the ability to have energy of motion. The higher you raise the ball, the greater its potential energy (and the more it will hurt when you drop it on your foot). Likewise, if you could pull an electron outward against the electromagnetic force binding it to the nucleus, you would also give it energy. Each orbit therefore has a different energy associated with it; the larger the orbital radius, the greater the energy. Because electrons are confined to specific orbits, they can take on only specific quantities, or "quanta," of energy (the theory is consequently called quantum mechanics). Each orbit therefore occupies its own "energy level." For any atom or ion there is always an orbit of smallest radius and lowest energy level called the ground state. Though they are restricted in the energies they may take, the electrons are free to change those energies by jumping back and forth between the allowed orbits or energy levels. (The Bohr model of the atom has been superseded by a view of quantum mechanics in which the electrons are represented by "wave functions." In this view the electrons are not hard little balls but exist everywhere along their waves, thus surrounding the nucleus. Nevertheless, the Bohr atom provides a convenient and intuitive

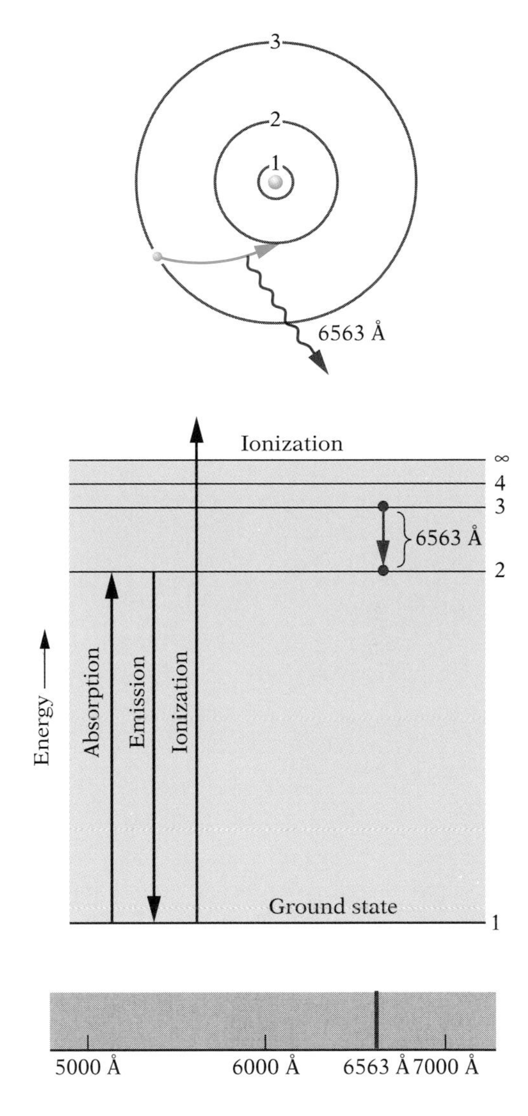

Electrons can exist only within allowed orbits, the first three of hydrogen shown above. Each has successively higher energy (middle). Electrons can absorb energy and be raised to higher orbits or energy levels (upward arrow), or they can jump downward with the emission of energy (downward arrow) in the form of photons. An electron jumping downward from orbit 3 to orbit 2 gets rid of its energy by emitting a photon at a wavelength of 6563 Å (bottom).

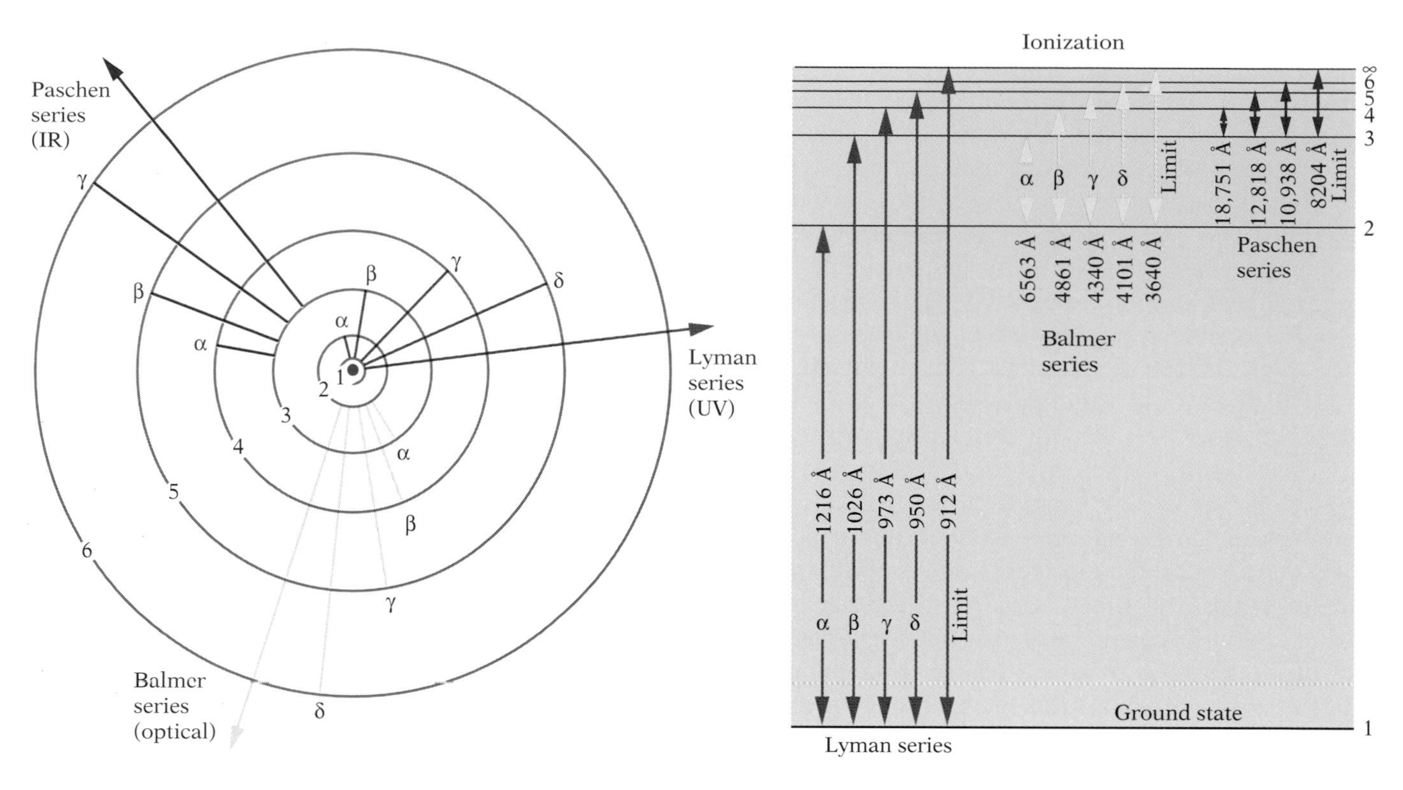

Hydrogen absorption lines (upward jumps) and emission lines (downward jumps) are arranged in series that have common lower orbits (left) or energy levels (right). The lines of the Lyman series (labeled α, β, γ . . . outward) arise from (or terminate on) the ground state, level 1, the Balmer series on level 2, the Paschen series on level 3, and so on. Since the energy jumps are greatest to and from level 1, the Lyman lines appear in the energetic ultraviolet part of the spectrum; several of the Balmer lines (Hα through Hϵ) appear in the optical; the Paschen glow in the infrared.

hook on which to hang valid explanations of the workings of the atom.)

If an "excited" electron, one in a higher energy level, jumps downward to a lower-energy one, it gives up energy. It does so with the release of electromagnetic energy by the emission of an energy-carrying photon. Since the electron's orbits have only specific amounts of energy, the *difference* in energy between the two orbits is also fixed, as is the energy of the emitted photon. And because the energy of a photon depends entirely on, and is directly proportional to, its frequency, hence inversely proportional to its wavelength, the emitted photon's frequency and wavelength are *also* fixed. Any given atom or ion in isolation can therefore emit *only* photons whose energies correspond to the entire array of energy differences between the energy levels. Because each kind of atom or ion has a different orbital or energy-level structure, all the different kinds of atoms and ions have discrete, but unique, emission spectra. (If the gas is very dense, that is, if the atoms are in very close proximity, not in isolation, a continuous spectrum will be produced.)

Absorption lines, the dark reversals of the emission lines seen against a continuous spectrum, are made by the reverse process.

Electrons can jump downward in energy spontaneously, but to jump upward to a higher energy level, the electron must get a kick—that is, it must absorb energy from an external source. One source of this energy is collisions among atoms: the energy of collision is transferred to an electron, causing it to jump to an outer orbit, from which it can jump back down to produce photons that will contribute to emission lines. The other possible source is electromagnetic radiation. An electron in any energy level can absorb a photon from radiation flowing past it; however, two conditions must be met. First, for a specific jump from one energy level up to another, the photon must have sufficient energy to take the electron all the way. Second, the photon cannot give up part of its energy but must yield *all* of it or *none* of it. As a result, only photons with energies exactly equal to the energy difference between two energy levels can be removed from the radiation flow. If continuous radiation streams through a gas, only photons of specific energies (and wavelengths) will be absorbed. The result is an absorption-line spectrum also unique to the kind of atom in the gas.

Hydrogen lines are arranged in series, named for their discoverers; each series has a common lower orbit or energy level. All the optical lines, those of the Balmer series, arise from or terminate on level 2, those of the Lyman series on level 1 (the ground state), those of the Paschen series on level 3, and so on. The biggest energy jump between adjacent levels is from 1 to 2; even though the orbital radii rapidly increase in size, the energy differences get smaller. As a result, the lines of a series get closer together as the number of the upper energy level increases, and the most energetic photons are associated with the Lyman series, followed by the Balmer, the Paschen, and the others. The Lyman lines therefore appear in the energetic short-wave ultraviolet part of the spectrum, the Balmer lines principally in the optical and the neighboring ultraviolet, the Paschen lines in the less energetic infrared, and so on.

The concept is beautifully illustrated—and, more important, supported—by the emission-line spectra of the diffuse and planetary nebulae. In the ultraviolet just below the optical, we see the Balmer lines marching to a limit at 3646 Å. Modern instruments have allowed us to see the infrared Paschen lines and those of even higher series. Radio telescopes now readily detect low-energy hydrogen emissions between *very* high energy levels, for example between levels 110 and 109 or 238 and 237. Similar spectra produced by neutral helium are also prominent in the optical, as are lines of oxygen, nitrogen, and neon, clear evidence that the nebulae are made of much the same stuff as found in the Sun and Earth.

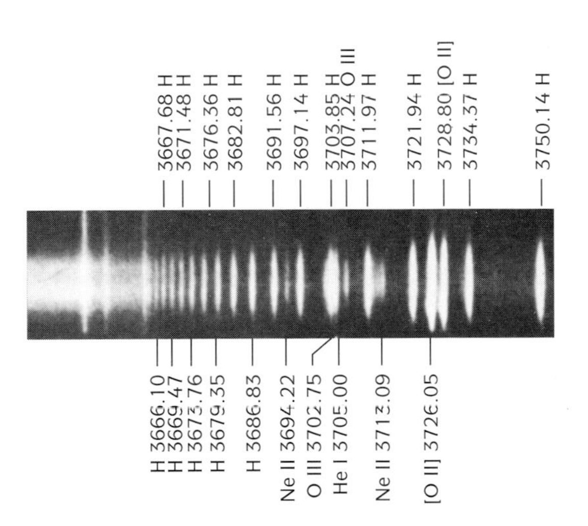

The hydrogen emission lines of the Balmer series (emitted here by a planetary nebula) get closer and closer into the ultraviolet, converging on a limit at 3646 Å.

SPECTRA AND ILLUMINATION

In 1927, the Dutch astronomer Hermann Zanstra applied the developing science of quantum mechanics to nebular spectra and discovered the primary illumination mechanism of the emission-line nebulae, both planetary and diffuse. For hydrogen emission lines to appear, hydrogen's electrons must somehow be boosted into outer levels. These higher energy levels are very unstable: any electron that finds its way into one will drop down to the inner, lowest-energy, level within about a 10 millionth (10^{-7}) of a second. Therefore, the vast majority of neutral hydrogen atoms must have their electrons in this lowest energy orbit, in the ground state.

All atoms and molecules can be ionized—stripped of one or more electrons—if they absorb photons with energies above a specific limit. A hydrogen atom with its one electron in the ground state can be "photoionized" by a photon with a wavelength less than 912 Å, a threshold at the end of the Lyman series called the Lyman limit. A star must have a temperature of at least 25,000 K to produce a significant number of such photons. The gaseous nebulae, diffuse or planetary, are ionized by stars at least this hot, a temperature that —consistent with the observations—corresponds to spectral class B1 (at the hot end of class B), thus explaining the tight relation between the O stars and the diffuse nebulae. Thus we have more than mere proximity to guide us: the O stars are *necessary* to the existence of the nebulae.

Zanstra first assumed his theoretical nebula to be "optically thick," completely opaque to ionizing radiation, radiation in the stellar continuum (continuous spectrum) shortward of the Lyman limit. Each of these extreme-ultraviolet stellar photons must therefore be absorbed by the gas, and each therefore produces one hydrogen ionization, that is, one free proton and one free electron. Eventually, the free electron will encounter a different free proton, recombining with it to produce a new neutral hydrogen atom. Each photoionization thus leads to one recapture or one recombination. A neutral hydrogen atom will be ionized quickly in the intense bath of stellar radiation. A free electron, however, can survive for months before encountering a recapturing proton. As a result, the nebular gas consists of a sea of free electrons and protons with only a few neutral atoms mixed in.

Look at the nebula from the point of view of the free electron. As the electron flies through the gas it sees the proton that will capture it surrounded by hundreds of steplike unoccupied possible orbits—energy levels—that lead downward to the ground state. The electron jumps in and can land on any one of the steps. There are no restrictions, except that the odds of going to the ground state are the highest and those of landing on successively upper states successively lower. If the electron is recaptured in the ground state, it simply recreates a photon in the high-energy Lyman continuum, that shortward of the Lyman limit. This photon ionizes another atom, it is as if the recapture never took place, and we start over until an electron lands on an upper level. All physical systems try to reach their lowest energies; you, for example, most likely are sitting down while you read a book. The electron will thus jump downward to any of the levels below it, creating an emission-line spectrum.

As an example, assume that the electron is recaptured on the fourth level. There are three lower levels, and the electron can jump to any of them with odds that can be calculated by the rules of quantum mechanics. Assume it jumps directly to the ground state, radiating a photon in the hydrogen Lyman series (here, Lyman γ, the third line in the Lyman series). If the gas is opaque to the Lyman continuum, it will also be opaque to the Lyman lines. As a result, the Lyman γ photon will immediately be absorbed by another neutral atom, raising its electron up to the fourth level, and again (statistically, in the community of atoms) it is as if no downward jump had taken place at all. Eventually the electron jumps to level 2 or level 3. If it goes to level 2, it will radiate Hβ at 4861 Å in the Balmer series (Huggins's third line). If the electron goes to level 3, it radiates Paschen α at 18,751 Å; the same reasoning shows that the electron must eventually jump to level 2 to create Hα at 6563 Å. The nebula is transparent to all hydrogen radiation except that of the Lyman lines or continuum, so these photons escape the nebula with no further interactions. Zanstra's important discovery was that in any possible combination of recaptures, the electron will wind up on the second level with the production of a Balmer photon. The electron will thence jump to level 1, radiating a Lyman α photon at 1216 Å, which bounces around until it reaches the edge of the nebula and escapes or is otherwise destroyed.

Zanstra thus showed that in an optically thick nebula, one that is opaque to ionizing radiation, the number of stellar UV photons produced per second

= the number of photoionizations in the nebula per second
= the number of recombinations in the nebula per second
= the number of Balmer photons radiated by the nebula per second
= the number of Lyman α photons generated in the nebula per second.

The result is that every stellar ultraviolet photon produces one escaping Balmer-line photon. We can calculate the odds of the recapture of electrons on different energy levels from theory and can therefore calculate the relative strengths of the Balmer lines. Some 80 percent of the Balmer photons come out of the nebula in the optical lines Hα through Hδ and, as a result, the nebula glows brightly to the eye. By far the strongest hydrogen emission produced in this "Zanstra mechanism" is the Hα line, which (unless the green 5007 Å line is quite strong, as it is in the Orion Nebula) generally gives diffuse nebulae their characteristic red colors.

The close agreement of the observed strengths of the Balmer lines with theoretical predictions provides powerful supporting evidence for the validity and precision of atomic theory (a benefit not only for astronomy but for other physical applications as well). This confidence in atomic behavior would eventually allow astronomers to use the observations of "recombination lines" to compute chemical compositions for comparison of nebular and stellar abundances, correspondences that tie the stars to the processes of the interstellar medium. Equally important, because the nebula converts ultraviolet radiation into optical radiation in a predictable way, it allows us to measure the radiative properties of stars in spectral domains that, because of the Earth's atmosphere, are inaccessible from the ground. The illuminating star thus helps us to understand the nebula, and the nebula in turn helps us understand both the illuminating O star and some of the fundamental processes of the atom.

Unlike bound electrons, those tied to the atoms, free electrons—those that run free in the gas after having been released through photoionization—can take on any energy at all. A free electron can approach a bare proton and change its energy *without* being captured—that is, it can go from one velocity to another, losing energy and giving off a photon. Any downward energy change is allowed, and therefore the collection of these "free–free" transitions produces a highly characteristic continuous spectrum. Free–free radiation is important because it makes diffuse nebulae shine quite brightly in the radio domain, and its unique spectrum allows these objects to be recognized for what they are even when they lie behind thick, obscuring clouds of what we now know to be star-forming dust. Since diffuse nebulae are markers of stars that have formed, we can locate regions of star birth that we cannot ever see with our optical telescopes.

THE MYSTERIOUS NEBULIUM

Although the nebular hydrogen lines were quickly identified, the bright pair found by Huggins at 4959 Å and 5007 Å remained defiant (as did the one at 3727 Å). Until it was seen that there were no stable elements left to be discovered, these lines were taken as evidence for an unidentified element called "nebulium" (after all, helium, named after Helios, god of the Sun, had been discovered in solar spectra before it was found on Earth). Continuing observations of planetary and diffuse nebulae through the early part of the century turned up many more of the mysterious unidentified lines, particularly strong ones in the ultraviolet at 3868 Å and in the red flanking Hα at 6548 Å and 6584 Å.

A year after Zanstra's pioneering work on the explanation of the hydrogen lines, Ira S. Bowen, then of the California Institute of Technology, put together the rest of the emission-line story in a classic example of the interaction between astronomy and the laboratory. Bowen, who later became the director of Mount Wilson and Palomar Observatories and developed powerful spectrographs for their large telescopes, was at the time exploring the laboratory ultraviolet spectra of the common ions. From the wavelengths of

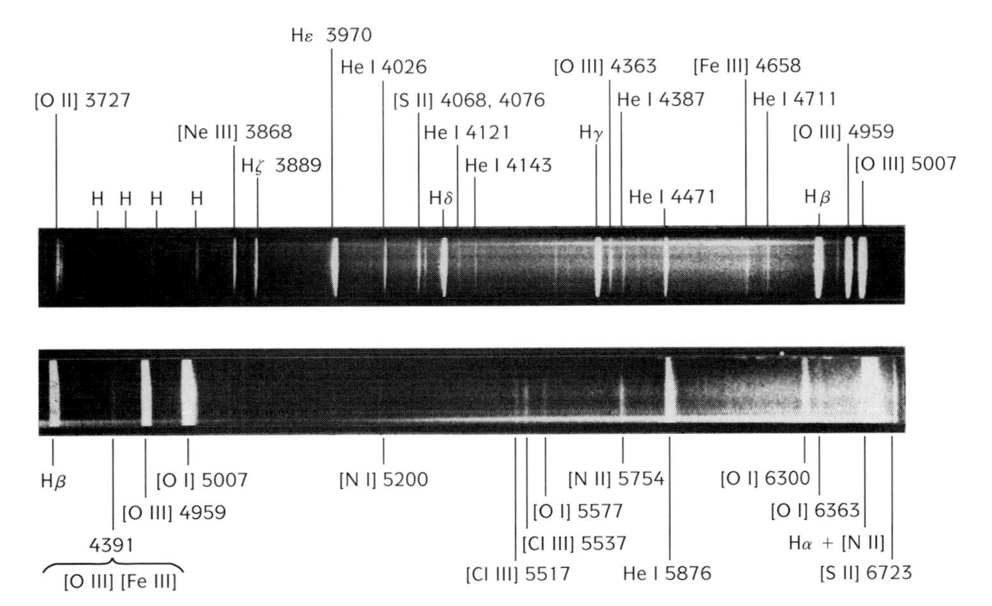

The spectrum of the Orion Nebula from the ultraviolet into the red reveals hosts of emission lines produced by a variety of elements. Roman numerals following the chemical symbols denote different ionization stages of the radiating atom: I for neutral, II for singly ionized, and so on. The brackets indicate "forbidden lines" that were unidentified until 1928.

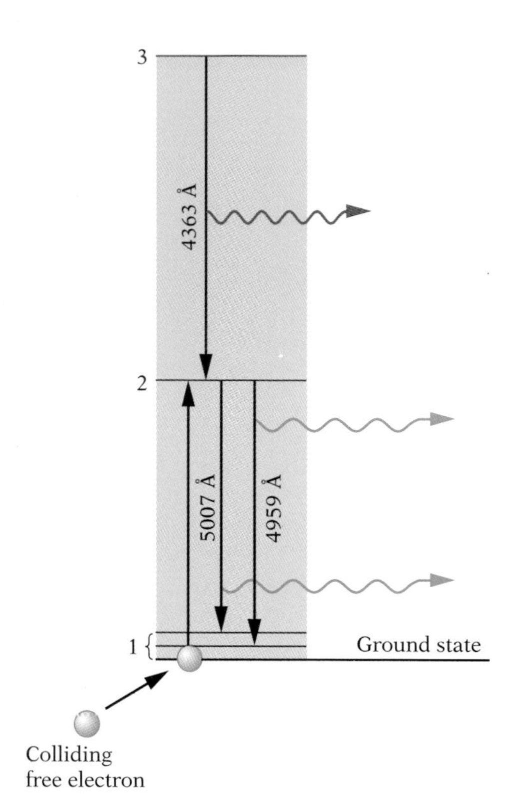

Collision with an energetic electron freed from hydrogen can kick doubly ionized oxygen's bound electron into one of a set of relatively stable excited states, from which the bound electron cascades downward to produce forbidden lines, including those first seen by Huggins at 4959 Å and 5007 Å.

observed ultraviolet lines, he could build a picture of atomic energy-level structures. He found that doubly ionized oxygen (oxygen with two electrons stripped away, expressed as O^{+2}) has a ground state that is split into three sublevels. There is also an energy level above the ground state that has energy differences relative to two of the sublevels that equal the energies of the photons of the nebulium lines at 4959 Å and 5007 Å. Bowen had found the source of the mysterious lines. By simplified rules of quantum mechanics, the transitions that produce them are not possible, hence are "forbidden." Only when we look at the complete mathematics of the atom do we see that these transitions are simply *unlikely,* the lifetime of an electron in the upper energy level measured in minutes rather than in millionths of a second. The great array of other unidentified emissions proved to be similar lines of diverse ions, the 3727 Å line produced by O^{+}, that at 3868 Å by Ne^{+2}, and those in the red at 6548 Å and 6584 Å by N^{+}. Bowen also identified additional "forbidden lines" (as they are still conveniently, if erroneously, called) of sulfur (especially prominent at 6717 Å and 6731 Å), argon, chlorine, iron, and a great number of others.

The forbidden lines are not produced by recombination but result from atomic collisions. The ionization of hydrogen from the ground state (where most electrons reside) requires a photon with a minimum wavelength of 912 Å. Almost all the ionizing ultraviolet photons will have shorter wavelengths and higher energies than needed. The freed electron therefore flies away with a speed that depends on the degree of the extra energy. If an electron freed from hydrogen has enough energy, it can slam into an O^{+2}, N^{+}, or other ion that has its *bound* electron in the ground state and lift that bound electron into an upper long-lived energy level, from which it eventually jumps back downward to produce a forbidden-line photon.

Ionic spectra are referred to by Roman numerals: that of neutral oxygen is O I, of the singly ionized state is O II, of the doubly ionized state is O III, and so on. Forbidden lines are indicated by brackets, so the 4959 Å and 5007 Å emissions of O^{+2} become [O III] lines, and that at 3727 Å is called [O II]. Several are identified in the spectrum of the Orion Nebula shown on the previous page. They cannot be seen in the laboratory because the long lifetimes of the levels that produce them render them weak. But under low-density nebular conditions, the recombination lines are even weaker and the forbidden lines dominate. Both kinds of lines are seen only because they are radiated by a vast amount of gas.

The forbidden lines, which add enormous richness to the entire spectrum, are observed from the ultraviolet far into the infrared.

For example, transitions that take place between the sublevels of the ground state of O^{2+} are seen in the infrared at wavelengths of 52 microns and 88 microns (a micron, μm, is 10^{-3} mm, or 10,000 Å). In the ultraviolet at 2332 Å and 2321 Å we see more [O III] forbidden lines that arise from a higher long-lived energy state above the one that generates the 4959 Å and 5007 Å lines.

Forbidden lines are not confined to the nebulae. We observe those of highly ionized metals in the spectrum of the hot outer layer of the Sun, the 2-million-K corona seen during solar eclipses, allowing us to measure a variety of coronal properties, one branch of celestial science aiding another.

OBSERVATION . . .

The emission spectra of diffuse nebulae reveal that they are made of warm, even hot, gases, but the spectra have not yet told us the conditions within the nebulae. For that we need the physical theories just described and a variety of quantitative measures made over a broad range of wavelength. Together, theory and observation allow us to learn temperatures, densities, chemical compositions, masses, mass distributions, and other properties of nebular matter.

Start with data collection. The astronomers from Huggins's time through the early twentieth century had to make do with eye-estimates of emission-line strengths as viewed either through the spectroscope or as represented on a photographic plate. While photographs are fine for recording appearances and structures, they are slow accumulators of radiation, even if carefully calibrated, and they are poor for making quantitative measurements of apparent brightness, as their response to light is unpredictable and not directly proportional to the amount of light falling on them. Only gradually did steadily improving electronic measurement take over, finally becoming the dominant technology in the past 30 or so years. Since the early 1970s, astronomers have settled almost exclusively on the charge-coupled device, or CCD, as the recorder of choice. A CCD plate consists of thousands of square cells that build up electrical charges when light falls on them. When an exposure is completed, the charges can be read out and the spectrum stored on disk or tape and reconstructed by computer on a video terminal. A CCD is a hundred times more sensitive than a photographic plate and has a directly proportional, linear response to light. As a result, astronomers can observe much fainter objects in much shorter periods of time and can accurately determine apparent magnitudes of stars and intensities of emission lines.

The techniques are not all that straightforward, however. A great deal of labor is often required to convert—"reduce"—the raw data into meaningful numbers. For example, in the optical spectrum the Earth's atmosphere absorbs light, and absorbs it better at short wavelengths than at long. Consequently, atmospheric absorption changes both the absolute and relative intensities of the optical emission lines. The wavelength dependence of the reflectivity of the telescope optics has the same effect. To account for these distortions, we must therefore also observe standard stars whose true spectral intensities as functions of wavelength are already known, itself a difficult and time-consuming task. The result, after calculating and applying the corrections, is the actual energy flow in each spectrum line at the distance of the Earth from the object observed.

Radiation in other wavelength domains can be similarly measured electronically. Radio telescopes are crucial to the examination of the interstellar medium, as they are used both to punch through the interstellar dust (which is transparent to radio waves) and to examine cold clouds that produce only low-energy radiation. A radio telescope uses a radio antenna in which an electrical current is excited by electromagnetic radiation from space. The current can be measured and the strength of the original signal found, allowing radio astronomers to measure the strengths of emission lines in the radio spectrum. And just as optical telescopes can image celestial sources, enabling astronomers to study structural characteristics, so can radio astronomers map the distribution of radio radiation across the sky to make radio images. Most important, in any spectral region we can combine the two kinds of observations, isolating spectrum lines with filters and taking pictures of our nebulae in optical light—or in the radio or infrared waves—radiated by specific atoms or ions to see how the different ions are distributed relative to one another

Because of absorption (and distortion) by the Earth's atmosphere, much of this work, especially in the infrared and ultraviolet, must be done from space. Space-based observations differ only in that the astronomer—or telescope operator—cannot have direct access to the instrument and must rely on radio commands. Whatever the mode of observation, the result is the data needed to find out what the nebulae (or any of the countless other astronomical objects) really are and how they are constructed.

. . . AND ANALYSIS

With theory and data in hand, it is possible to learn about the internal natures of the nebulae (and indeed of other cosmic clouds). As a result of the constant collisions between the free electrons, the electrons take on a particular distribution in velocities mathematically described by James Clerk Maxwell in 1860. In any kind of hot, opaque, dense gas—like that in the Sun—temperature describes the amount of energy radiated (through the blackbody laws), the velocities of the atoms and ions, the number of atoms with electrons in particular energy levels, and for a given element, the numbers of ions relative to the number of neutrals. A low-density gas like that in a nebula, however, which radiates emission lines, no longer behaves like a blackbody, in part because the gas is transparent and radiation does not couple well with its atoms. Temperature no longer dictates luminosity or level populations, but only the velocities of the particles, and in this restricted sense the temperature is called a kinetic temperature. Since in nebulae we deal with the velocities of free electrons, the kinetic temperature is an *electron* temperature.

The free electrons are heated, that is, given their initial velocities, by the exciting star. The hotter the ionizing star, the more energetic its radiated photons, the faster the nebula's free electrons after ionization, and the hotter the nebular gas. The electrons are cooled largely by giving up their energies in the collisions with heavy ions to produce the forbidden lines. The greater the number of heavy elements, the cooler the gas. The resulting electron temperature derives from the balance between the heating and cooling mechanisms.

The forbidden lines are crucial to the diagnosis of nebular conditions. Because the forbidden lines all arise from energy levels at different energies above the ground state, their relative intensities must depend not only on the relative ionic abundances but also on the relative energies of the free electrons. Doubly ionized oxygen has another long-lived energy level, above those that produce the 4959 Å and 5007 Å lines, from which arises a forbidden line at 4363 Å. In a cool gas, the electrons might have enough energy to excite the 5007 Å line but not the 4363 Å line. As temperature rises, the strength of the 4363 Å line increases relative to the 5007 Å line. The ratio of the lines therefore gives us the electron temperature. On the other hand, the 3727 Å line of [O II] actually consists of two lines only 2 Å apart. Their intensity ratio (and those of similar [S II] and other lines) is sensitive to density instead of temperature, allowing us to find the number of electrons per cubic centimeter.

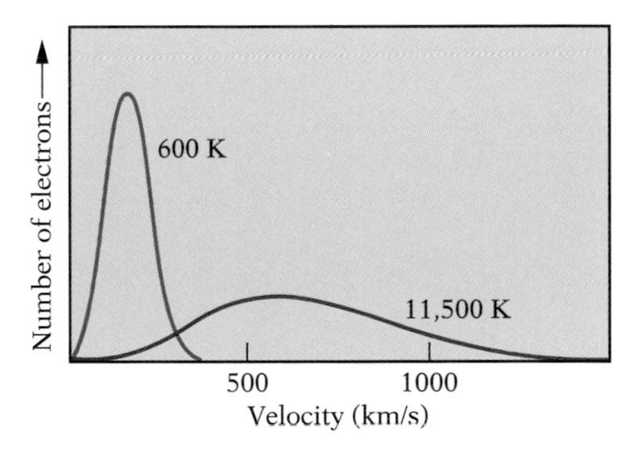

The velocity of electrons in a nebula depends on temperature, which can be found by sampling the velocities at two points or from the relative numbers of electrons above two specific energies or velocities (determinable from pairs of forbidden lines of the same ion).

The Orion Nebula has an electron temperature of about 9000 K (depending on the location in the nebula) and a density of about 10^4 electrons cm^{-3} in the center, dropping to below 10^2 cm^{-3} at the edge. The central density, actually high for a classic diffuse nebula, is comparable to the best vacuum that can be produced on Earth. As a result, the nebulae let us examine physical processes under conditions impossible anywhere else, processes that cannot be studied in any laboratory, in this way leading us to a more complete knowledge of the atom and its spectrum, and of nature itself. A cubic meter of a nebular gas would radiate next to nothing; nebulae are visible only because of their large masses, which can be found from the average density and the volume. The mass of the Orion Nebula is roughly 500 times that of the Sun; that of the gigantic Tarantula Nebula is over 100,000 times the solar mass.

Once we know the excitation mechanisms of the recombination and forbidden lines, and have acquired data from all across the spectrum, we can calculate the relative electron populations of the energy levels, which depend on temperature and density. We can then relate the relative strengths of the lines to relative ionic abundances and finally calculate chemical compositions. The compositions of the diffuse nebulae are satisfyingly close to that of the Sun and the local stars (for every 10,000 hydrogen atoms, we find about 800 of helium, 4 of oxygen, 0.6 of nitrogen, about 3 of carbon, and 2 of neon). Children have the characteristics of the parents; the similarity in chemical compositions between the diffuse nebulae and the stars more firmly than ever ties the nebulae and the interstellar medium, of which they are a part, to their progeny, the stars.

Bubbles in the Medium

Nebulae are dynamic affairs that strongly affect their surroundings. Light up a hot star in a large cloud of neutral hydrogen. Every ionizing ultraviolet photon will eventually be absorbed by the neutral gas, which will become nearly fully ionized out to a radius, an "ionization front," at which all the ultraviolet photons are absorbed. A diffuse nebula set in a uniform neutral medium will therefore be a sphere in which the ionized gas becomes neutral at a sharp boundary. This idealization is called a Strömgren sphere (after the astronomer Bengt Strömgren, who developed the concept in 1939) and has a radius that depends on the gas density and on the number of ionizing photons, which in turn depends on the number of hot stars and their luminosities and temperatures.

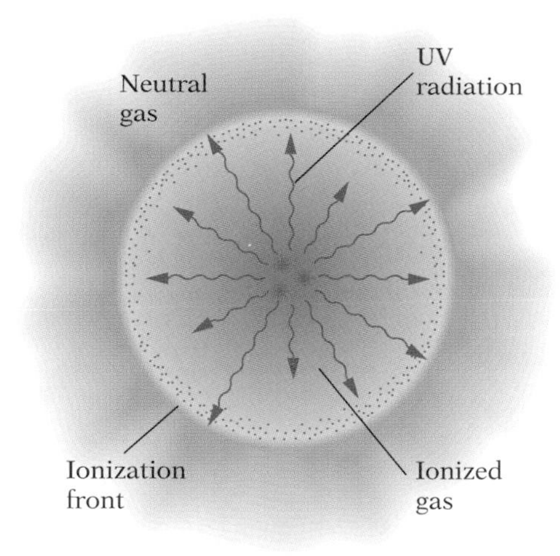

The ideal diffuse nebula is a Strömgren sphere, an ionized cloud set within a neutral medium. The star or stars ionize the nebula out to a radius—the ionization front—at which all the ionizing ultraviolet photons are absorbed. The Rosette Nebula is an excellent approximation to the theoretical ideal.

When the ionizing star is suddenly switched on, its photons quickly eat into the surrounding gas, carving out an equilibrium Strömgren sphere in which the total number of ionizations per second exactly equals the number of recombinations per second. The ionization heats the gas, which increases the gas pressure above that in the neutral surroundings, and the bubble expands even faster, aided by the lowered density, which allows greater penetration of photons. The sphere then rapidly grows, eating into its neutral surroundings, perhaps to the point of bursting out of its interstellar cocoon.

The classic example of a Strömgren sphere is the nearly circular Rosette Nebula in the obscure constellation Monoceros, east of Orion. Other diffuse nebulae can be highly distorted as a result of irregular mass distributions. In some instances, the nebula is not optically thick (absorbing all incident ultraviolet photons) but *optically thin;* that is, some ultraviolet photons penetrate the cloud and

An expanded Hubble view of the whole Orion Nebula, a blister on the edge of a dark neutral cloud, shows beautiful curls of gas that fountain out of the center toward the observer. In the middle is the bright central core imaged by Hubble. Much of the delicate structure is caused by shock waves. To the north, directly above the Orion Nebula, is the much fainter M 43.

escape into space. In that case, we see something of the actual shape of the cloud itself; alternatively, the stars may not be centered within a cloud, or the cloud might be optically thick in one direction and thin in another. Though the Orion Nebula looks vaguely like a Strömgren sphere, it is actually a complex blister on the edge of a vast neutral cloud. Other distortions are caused by multiple ionizing centers, or merely by variations in gas density that produce nonspherical ionization fronts.

More detailed looks at the Orion Nebula show increasing complexity, the gas broken into sheets and filaments. This extraordinary and beautiful network results from a three-dimensional structure with curved and irregular ionization fronts projected onto the two-dimensional sky, from turbulence within the gas, and from embedded stars. Even a gas as tenuous as this one, with a density of only a few thousands of atoms per cubic centimeter, will conduct sound waves (with a speed in the neutral medium of a few kilometers per second). If the gas is forced to move at a greater speed, it piles up into a shock wave, a moving wall of pressure. The most familiar examples are the bow wave of a boat moving quickly through water and the sonic boom of an aircraft flying at supersonic speed. (The sonic boom does not take place, as commonly believed, only when an airplane "breaks" the sound barrier. It is a continuously moving shock generated by a plane *already* flying above sound speed. We hear the sharp "thump" as the shock passes by us.) The push of the ionization fronts generate shocks—sonic booms—that propagate within the cloud, helping to break it into a delicate, foamy tracery and that excite more atoms to produce more emission lines with a characteristic spectrum that identifies the shocks.

The beauty of the diffuse nebulae arises first from the ionization of the gas by embedded O stars, recombination and collision producing the emission lines that give the nebulae their colors, the ionization boundaries and shocks subsequently fragmenting the nebulae into their characteristic structures. The appearances of the nebulae, however, are also powerfully influenced by billows of dust that hide some of the illuminated material, dust that both surrounds and penetrates the nebulae. The bright clouds highlight the dust and show us where star formation has already taken place. We now dive into the darkness of the dust, probing ever deeper until we can see the stars being born, some of them destined to give future nebulae their colorful glory.

3

DARK DUST

≺ *Black motes of thick dust spatter against IC 2944, a bright nebula.*

"*I*n my late observations on nebulae," wrote William Herschel in 1784, "I soon found that . . . the spaces preceding them were generally quite deprived of their stars. . . . More than once, I ventured to give notice to my assistant at the clock to prepare [to record nebular positions], since I expected in a few minutes to come at a stratum of the nebulae, finding myself already (as I then figuratively expressed it) on nebulous ground." This was the first scientific mention of "stellar voids" in the Milky Way. Herschel's words were a visionary intimation that these dark regions were in fact *not* "stellar voids" but somehow nebular in nature.

Though the broad spatial relations between these bright and dark regions were clear to early astronomers, knowledge of their physical relationships, and their links to the formation of stars, had to wait for the technology of our own time.

Dark constellations of the Milky Way, here centered on the Coalsack to the left of the Southern Cross, were created by the Incas of South America.

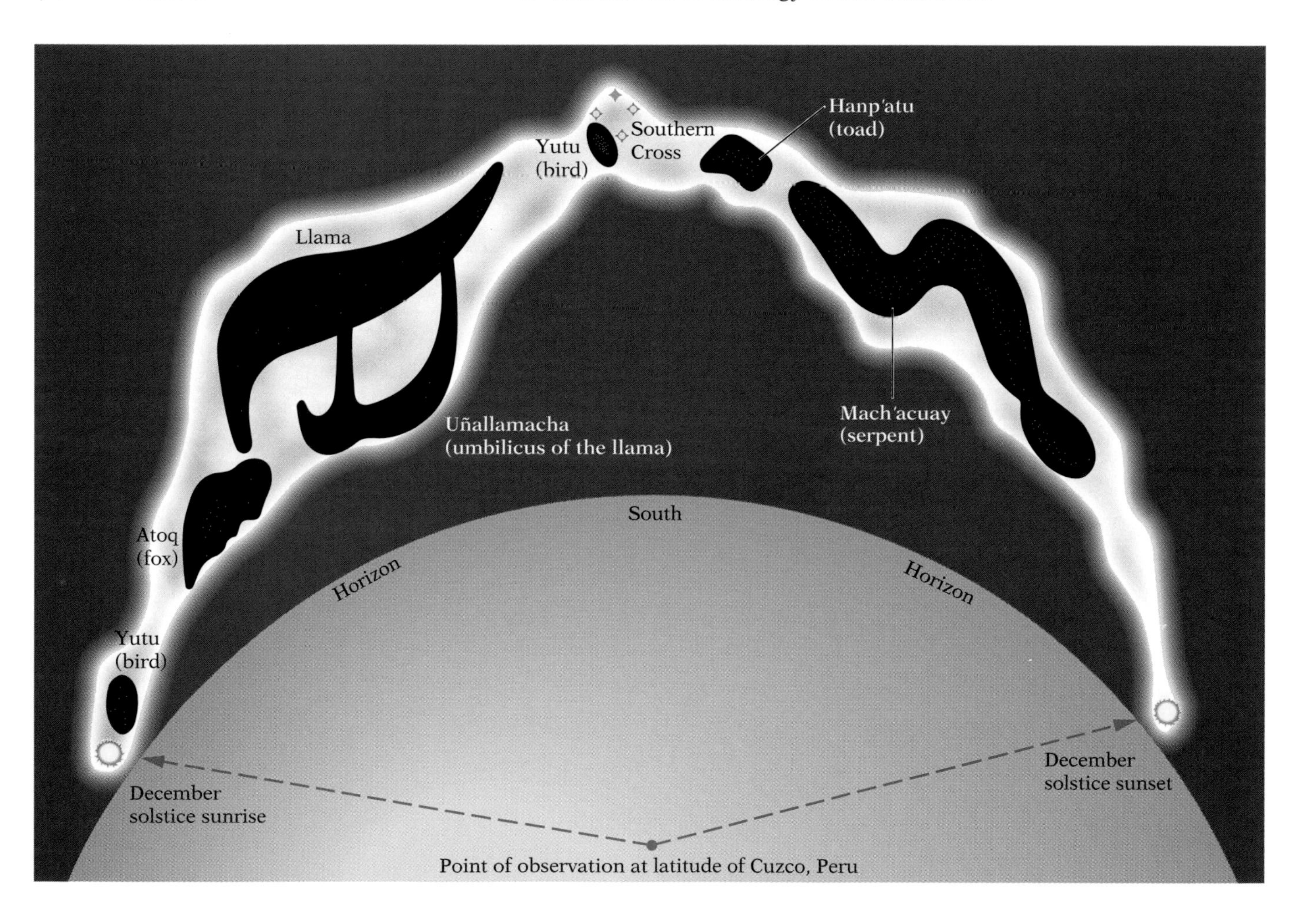

Shadows in the Night

The shadowy clouds of the Milky Way are everywhere, piled one atop another against the wide, graceful trail made of the billions of stars of the galactic disk. Above Peru and Chile, the yawning gaps in the central Milky Way are so apparent that the Incas gave them names. These "dark constellations" are part of an intricate black lane that splits the brightest part of the Milky Way, between the Southern and Northern Crosses, neatly in two; the northern portion of the lane is known as the Great Rift. A deeper view shows that the Rift and its dramatic southern counterpart are actually great numbers of dark cloud complexes (like the obvious Coalsack, in the Southern Cross, or Crux, which is the Incas' Yutu). Clouds like these also riddle the entire circle of the Milky Way and are most apparent where the background light is brightest.

The Coalsack and similar complexes are made of a number of discrete blobs called globules, or sometimes Bok globules, after Bart Bok, the eminent Dutch-American astronomer who brought them to prominent attention. Individual globules are also seen as isolated patches, a degree or a few minutes of arc across, nestled among the stars of the Milky Way. Some are just barely detectable as slightly diminished areas of stars; others are absolutely black, totally opaque to any background radiation. Most globules have vaguely

With the center of the Galaxy in the zenith, the Milky Way cascades to both horizons: the Northern Cross is to the left, the Southern to the right. Hundreds of dark clouds combine, splitting the Milky Way in two.

Bright and dark clouds form the Horsehead Nebula in Orion (above); a close-up of the Rosette Nebula (at right) shows "elephant trunks" and tiny specks.

circular outlines (and roughly spherical shapes), but a few, like the prominent Snake in the constellation Ophiuchus, are stretched into filaments. Others point like fingers toward nearby bright stars.

The smaller they are, the more intriguing and prominent the dark clouds become. Below a certain limit, they can no longer be easily detected against the stars, but if they are backlit by an associated diffuse nebula, even the details and intricacies of their structures can be seen. The Orion Nebula displays two powerfully contrasting features, the "bright bar," an ionization front viewed nearly edge-on, and the "thumbprint," a dark intrusion from the north, beautifully defined in G. P. Bond's drawing, that looks like a nebular bug squashed by some celestial Zeus. This apparent hole in the background brightness is actually caused by a cloud of dark obscuring matter positioned in the line of sight against the glowing gas. Close examination of any diffuse nebula quickly reveals many such features that give the nebulae much of their complexity: the "bay" of the Lagoon Nebula, the "Gulf of Mexico" in the North America Nebula. Like the Incas, we have named the most prominent of the dark clouds, the Horsehead Nebula in Orion perhaps the most dramatic. Against numerous nebulae we also see striking long "elephant trunks" and impenetrable black specks that seem to have been spattered by some mad abstractionist.

Early observation merely showed dark islands set within the river of stars. What were they? Modern scientific scrutiny of the question really began around 1910 with the work of Edward Emerson Barnard. With a 10-inch lens at Yerkes Observatory, he photographed the bright nebulosities around the star ρ Ophiuchi, which are connected by a complex set of dark lanes and clouds, and those around its neighbor, ν Scorpii, 5° to the north. The close relation between the dark and bright regions first noted by Herschel strongly suggested to Barnard that the dark patches were caused by masses of obscuring matter. Particularly important was his discovery that the bright edge of a small diffuse nebula to the west and north of ν Scorpii was contiguous with the dark edge of the apparent "hole," certainly a remarkable coincidence if the dark area were merely a gap in the stellar distribution. By 1919, when he published a catalogue of 182 dark objects (extended in 1927 to 370), Barnard was convinced of his opinion. The more prominent dark clouds still carry his designations: the Snake is known as B 72, the darkest lane near ρ Ophiuchi as B 44.

The deeper we look into the Milky Way the more dark clouds we see. In the 1950s, the entire sky visible from Palomar Mountain in southern California was photographed in 5°-wide blocks in two colors (with red and blue filters) with the 48-inch Schmidt telescope. B. T. Lynds, then of the National Radio Astronomy Observatory, used this *Palomar Sky Survey* in 1962 to catalogue more than 1800 dark nebulae (thus the Snake, B 72, is also L 1495). Bart Bok estimated that there may be 25,000 observable globules, and there are probably many times more.

Other evidence shows the physical reality of the dark clouds. As they orbit the galactic center, stars are in continuous motion relative to one another. If the dark nebulae were truly empty, if they were holes in the stellar distribution, stellar motions would quickly blur the sharp edges of the dark areas unless some mysterious physical process were keeping the stars at bay. The bright nebulosities that enfold some globules also reveal the presence of real, though dark, objects. Furthermore, a few stars seen through the more translucent of the globules are anomalously red, demonstrating that the globules are somehow affecting the starlight. Something within them selectively removes the blue component, much as Earth's

Bart Bok, one of the luminaries of twentieth-century astronomy, made the Milky Way a lifetime study.

E. E. Barnard at work wearing his reindeer-skin coat.

atmosphere reddens the setting Sun. Additional evidence is provided by other galaxies: those with disks presented edge-on show us dark lanes running down their spines, exactly what is observed in our own Galaxy. If the dark lanes were gaps, in time they would close up because of the parent galaxy's gravity.

The vast lane of obscuration in the Milky Way, composed of the set of dark clouds, is the narrowest feature of our Galaxy's disk. Along with the O and B stars—and the diffuse nebulae that surround them—it defines the central plane. The link between the newly formed O stars and the diffuse nebulae, and now that between the nebulae and the dark clouds, make a chain that shows the relation between the dark clouds and young stars. In 1922, in a burst of perspicacity, Henry Norris Russell, and later Bart Bok, called attention to the likelihood that stars are actually born within the dark clouds. When a new star is sufficiently massive and hot (an O or a hot B star), it illuminates the cloud to produce a diffuse nebula. Since the diffuse nebulae are gaseous, the dark clouds must be mixtures of gas and obscuring matter. The relations among the various kinds of dark clouds and the way they evolve involve—at least in part—the natures of the surrounding stars and star formation itself.

Taking the Measure

In the early 1920s, the German astronomer Max Wolf played a major role in the intellectual illumination of the dark nebulae. Not only did he demonstrate, by an essentially statistical method, that light is absorbed in the dark areas of interstellar space (thus establishing that those areas are material), he also devised a way to measure the distances of the dark clouds. He observed that the number of stars per square degree to the west of the Veil Nebula in Cygnus (which is the remnant of an exploded star) is very noticeably less than the number to the east. Something apparently related to the nebula is dimming the stars, hiding the fainter ones. Wolf simply counted stars per unit apparent-magnitude interval within the obscured region and compared the results to the counts in unobscured areas. The numbers of 9th- and 10th-magnitude stars per unit angular area were about the same in both the obscured and unobscured regions; but in the obscured region the number of stars per unit area of 12th magnitude and fainter was significantly smaller. The cloud that causes the light obscuration must therefore be at about the distance of the average 11th-magnitude star.

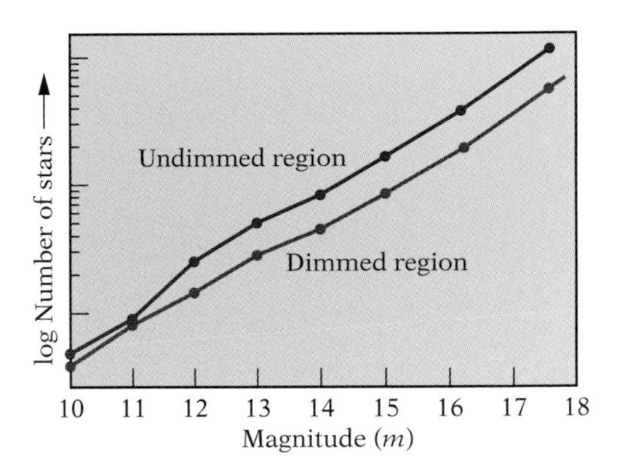

The Veil Nebula in Cygnus, the ancient remnant of an exploded star, divides an area rich in stars from one made distinctly poorer by a sheet of obscuring matter. Stars in front of the cloud are undimmed; those behind are all dimmed by a similar amount. The graph of star counts shows that the number of stars fainter than 11th magnitude is diminished by the dark cloud.

The problem faced in this simple analysis is that stars have a huge range in intrinsic luminosities, so it is a tricky statistical problem to determine the distance of an "average" star of 11th magnitude. From a knowledge of the distribution of stars as a function of luminosity, Wolf estimated that the cloud was about 1600 pc away. Since there appears to be a physical relationship between the Veil Nebula and the dark cloud—each is at the edge of the other—the Veil is at about the same distance. Such star counts also showed that the globules that make up the Coalsack are about 150 pc away.

The Coalsack in the Southern Cross, among the darkest of the larger dark cloud complexes, is roughly 150 pc away; it almost completely obscures the background stars.

From its 4° angular diameter, the Coalsack must be about 10 pc in diameter, eight times the distance between the Sun and α Centauri.

Wolf's technique, however, is only approximate, and it serves to place obscuring clouds only as long as a sufficient number of stars can be seen in front of them. Much more precise distances can be derived if the dark cloud is clearly associated with a diffuse nebula whose exciting—ionizing—stars can be recognized. Because the stars are within the bright nebula, we need only find the spectroscopic distance of the stars to find that of the whole complex. However, the obscuration may spread beyond the boundaries of the obvious dark clouds and may therefore dim the starlight, making the embedded stars appear to be farther away than they are. We need to know more about the nature of the obscuration and how to correct for its effects. Such discoveries will also give us clues to how the stars are born within the clouds.

MIRRORING THE STARS

In the early part of the twentieth century, Vesto M. Slipher, an astronomer at the Lowell Observatory in Arizona, mounted a project to observe the spectra of the nebulae and made two discoveries that helped separate them into their modern categories. In 1917, he showed that the spectrum lines of most spiral nebulae (which are now known to be galaxies) are shifted toward the red part of the spectrum; only M 31 and its neighbor M 33 had blue shifts. He logically interpreted his result in terms of the Doppler effect, a change in observed wavelength. If a body emitting a wave approaches you (or you approach the body, or both), you hit the waves more frequently and the wavelength—the distance between the crests—appears smaller than if the body (or you) were at rest; if you and the emitting body are moving apart, the measured wavelength is greater. The shift in wavelength is directly proportional to the speed along the line of sight, the radial velocity. You can easily see the effect by rowing a boat into, and then with, water waves.

By showing that the spiral nebulae are moving away from us, the redshifts were the first real evidence that these structures might be outside our own Galaxy. In fact, they are large external stellar systems, as later established by Edwin Hubble when he began measuring their distances in 1928. Hubble then found that the redshifts correlate with distance, observationally revealing the expansion of the Universe. We know now that the systematic redshifts found by Slipher are not the result of the Doppler effect but of the expansion of space itself. Nevertheless, Slipher's work prepared the way for the formulation of the modern theory of the Universe, the Big Bang: that the Universe began in a high-density, high-temperature state some 15 billion years ago and has expanded steadily since.

Vesto Slipher is also known for his discrimination, in its way of equal value, of two kinds of galactic nebulae (those, like the diffuse nebulae, that lie *within* galaxies). William Huggins had found that the Orion Nebula radiated emission lines. In 1913, Slipher discovered that much fainter wisps of nebulosity surrounding Merope, the brightest star in the Pleiades star cluster, showed *absorption* lines in a spectrum that was an exact copy of the star's. The nebula was reflecting the stellar spectrum. Ejnar Hertzsprung immediately suggested that the starlight was being reflected by solid particles or grains of matter, what we now (once their dimensions were known) simply call *dust*.

We owe the explanation of the difference between the reflection nebulae and the diffuse nebulae to Hubble. Before his seminal

Vesto Slipher, a Lowell Observatory astronomer, helped discriminate between reflection and diffuse nebulae and laid observational groundwork for the discovery of the expansion of the Universe.

Bluish reflection nebulae in Cygnus, illuminated by stars set within dusty clouds, contrast beautifully with the reddish, ionized diffuse nebula NGC 6914.

discoveries of the nature of galaxies and the expanding Universe, he found that the nebulae, whether emission or reflection, were intimately related to their central stars: the brighter the star, the larger the nebula, showing that the buried star is indeed responsible for the nebular illumination. He also found that the emission and reflection nebulae divided neatly along lines of stellar spectral class. The emission nebulae are excited by stars of 25,000 K or hotter (the B0 stars and all the blue O stars); the reflection nebulae, however, are illuminated by stars of class B2—20,000 K—or cooler. The nebulae associated with class B1 are an equal mixture of emission and reflection.

Hubble's work prompted Henry Norris Russell to relate reflection nebulae to dark nebulae, declaring that "it appears probable that the aggregate mass contained in one of these great obscuring clouds must be very considerable—probably sufficient to form hundreds of stars—and that a sensible fraction of the whole mass must be in the form of dust less than 0.1 mm in diameter."

The faint continuum of the Orion Nebula was subsequently also found to contain, as well as bright emission lines, absorption lines that are a weak but faithful copy of those of θ^1 Orionis C, the principal exciting star: the bright gas of the diffuse nebulae is also mixed with dust. Finally, the picture was complete. The gas in interstellar space is thoroughly mixed with solid grains—with dust. If there are no nearby or embedded stars, a cloud of this matter is dark, the dust blocking the background starlight. If there is a hot embedded star (or stars), hotter than class B2, it radiates sufficient ultraviolet photons to ionize the gas, and the resulting emission-line spectrum is so bright that the faint reflection spectrum produced by the dust is generally hidden. If the star is class B2 or cooler, the gas remains neutral, and a reflection nebula is seen.

The dark clouds are differentiated from the reflection nebulae by physical processes as well as by appearance. Place a star behind a thick dusty cloud. Some of the radiation that passes into the cloud is absorbed, actually stopped, some grains taking the radiation into, and heating, themselves. Other photons are scattered, bouncing off individual grains. Although the photons escape the cloud, they

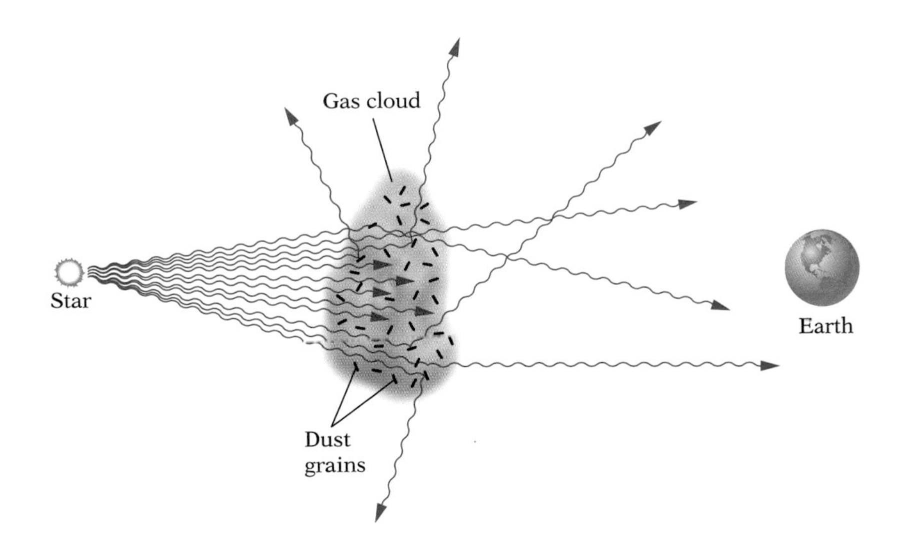

Some photons from a star behind an interstellar cloud are absorbed, some scattered, by the dust grains in the cloud. Scattered light produces a reflection nebula. If the cloud is thick enough, no light gets through and the cloud is seen as a globule.

change direction, so the amount of light along the line of sight is much diminished. The probability of a photon's being either scattered or absorbed is the "extinction." The relative importance of the two phenomena, scattering and absorption, is expressed by the albedo, here the ratio of the probability of scattering by a particle to the total extinction.

The darkness of a dark nebula depends on the number of grains along the line of sight to the background stars (the density of grains cm^{-3} times the path length) and the extinction per grain. A thick cloud is seen as a dark nebula, but because of the scattering, a thinner cloud may glow slightly or even appear as a reflection nebula. The brightness of a reflection nebula depends on the amount of scattered light as well as on the relative efficiencies of scattering in different directions. Hubble's comparisons of the reflection nebulae and their illuminating stars showed that the grains are rather bright. Computer modeling of the nebulae indicates average albedos of about 0.6 and shows that the grains have a strong tendency to scatter light in the forward direction (the direction the light was initially going). As a result, reflection nebulae that have a star behind them are brighter than are those that have stars in front. Much the same phenomenon is seen in the daytime sky: high translucent cirrus clouds are much brighter when the Sun is behind them than when it is to the side because of enhanced forward scattering. Moreover, the albedo obviously climbs with decreasing wavelength:

This dark cloud in Scorpius is clearly affected by the nearby star, demonstrating the presence of a dark physical object.

reflection nebulae, in stark contrast to the reddish diffuse nebulae, invariably appear blue, the effect enhanced by the bluish color of the illuminating B stars. The extinction and the albedo depend mostly on two variables, grain size and chemical composition.

INTERSTELLAR EXTINCTION AND REDDENING

Since dark clouds and globules were recognized everywhere in the Milky Way since Herschel's time, their importance to large-scale studies of the Galaxy should probably have been recognized earlier than it was. Though the potential importance of a pervasive extinction in the Galaxy had been argued, many astronomers were perhaps "in denial," insisting that except for the discrete and obvious clouds, space was largely clear. In 1930, however, R. J. Trumpler of Lick Observatory found persuasive evidence for visually *undetectable* regions of obscuration, obscuration that might affect the whole galactic plane. Trumpler thereby launched astronomers on a difficult yet ultimately satisfying journey in their quest of accurate distance.

Trumpler had been engaged in studying the properties of open clusters. The first step in estimating the distance to an open cluster is finding what kinds of stars the cluster contains by establishing its HR diagram, plotting apparent magnitude against spectral class. Even in 1930, Trumpler had reasonable estimates for the absolute magnitudes of the different kinds of stars. He could then apply the magnitude equation to each of them; the average solution was the cluster's distance. Knowledge of the distance and the cluster's angular diameter allowed derivation of the physical diameter, which is typically 10 pc or so.

Trumpler found that the more distant clusters in his sample were twice as large as the closer ones, no matter in what direction he looked. When you observe a phenomenon in the Galaxy that seems to have the Earth as its center, you know something is amiss. Trumpler added up the evidence and realized that there must be an obscuring medium between us and the distant clusters, in fact, between us and all the clusters, that makes the more distant stars seem fainter than they actually are. The result is a systematic overestimation of distance. To produce this effect the dust had to be everywhere, in a quasi-homogeneous, lumpy medium into which is set a huge variety of denser clouds. As a correction, Trumpler suggested a more or less uniform dimming, or extinction, of about 0.7 magnitude per kiloparsec of distance. The discovery had

NGC 891 is a galaxy similar to ours; its lane of obscuring matter is clearly seen because the galaxy is edge-on to us. Note the great similarity with the rift that divides the Milky Way.

far-reaching effects: it explains why distant galaxies are never seen in the plane of the Milky Way and why our Galaxy seems to be so much larger than the others. Imagining ourselves inside NGC 891, seen edge-on in the photograph above, we can easily understand how its residents would see the same effect.

To apply the method of spectroscopic distances (from which we found the distances of the diffuse nebulae and open clusters) we must correct the apparent magnitude of a star for the dimming by dust. This means knowing the degree of extinction (A), the amount by which the apparent visual magnitude must be brightened (that is, the amount by which its value must be decreased) before it is entered into the magnitude equation. How can we possibly know A? Trumpler again showed the way. The stars of his more distant clusters are reddened as well as dimmed. The color of a star, as defined in astronomy, is found by measuring its magnitude at two wavelengths. The eye sees best in the yellow part of the spectrum, at a wavelength of about 5500 Å, and traditional magnitudes, technically called visual magnitudes, are tied to that wavelength. But what would the sky look like if we could use a *different* wavelength? The early photographers saw the effect right away: photographic plates are much more sensitive to blue and violet wavelengths than they are to yellow. That sensitivity, combined with the transmission characteristics of the Earth's atmosphere, means that photography sees best at 4500 Å in the deep blue. Photographic filters that matched the response of the human eye were used to define visual magnitudes.

Stars behave like blackbodies, so cool stars radiate more at longer wavelengths than at shorter, hot stars at shorter than at longer. A photographic magnitude system (m_{ptg}) sees hot stars as brighter than does the visual-magnitude system (m_{vis}), and cool stars as fainter. Color can therefore be quantified in a "color index" by taking the difference $m_{ptg} - m_{vis}$. Photography is no longer much used in professional astronomy, and the two standards have been replaced by electronically recorded versions at the same wavelengths called B (for blue) and V (for visual). Color index is thus $B - V$. B and V are set equal for white stars of class A. The color index therefore ranges from –0.4 for the hottest stars (class O) to about +2 for class M. Since spectral class correlates with temperature, it also correlates with $B - V$; the relation is well established from observations of nearby stars that suffer insignificant interstellar extinction.

We can find A (properly A_V if we are working in the standard visual system) because it scales with the degree of reddening. Reddening (sometimes called selective extinction to distinguish it from A_V, the total extinction) is quantified by the color excess (E), the difference between the value of $B - V$ we actually see and that which we *would* see were the star not reddened. Since reddening produces higher color indices, E is always positive.

The trick is to find the relation between E and A_V. Years of examining a variety of stars have shown the ratio $R = A_V/E$ to be reasonably constant from one place in the Galaxy to another, and astronomers speak of a "universal reddening law," with R equal to about 3 (the standard value is 3.2). By observing a star's color, we can derive the intrinsic color from the spectral class, which gives us E, then $A_V = RE$, which allows the correction of V—the apparent magnitude—and thus the true distance.

The only difficulty is that the value of R is dependent on specific dust properties, and in some cases (as toward the Orion Nebula), R is *not* equal to 3, so we can make inadvertent errors. Walter Baade, the Mount Wilson–Palomar astronomer who first differentiated the halo and disk populations of stars in galaxies, was once asked if he would become an astronomer if he had to do it over again. "Only if I could be assured that the ratio of total to selective extinction were the same everywhere," he replied.

A Microscope on the Grains

The nature of interstellar reddening tells us a great deal about the properties of the grains. If a photon has a long wavelength, long relative to the size of the grain, it will not "see"—intercept—the grain; rather it will pass right by, unaffected. The closer the wavelength to the grain dimension, the more likely the two will interact. If you are in a rowboat at sea, you do not notice long swells, but if the waves have a length similar to the boat's dimension, you may be swamped. Similarly, on a foggy day in which minute droplets of water absorb light, you can listen to your local radio station with no diminution of signal.

The most familiar example of the phenomenon is the blue sky, which results from the interaction of sunlight and the Earth's atmosphere. While the long radio waves penetrate the air quite well, the wavelengths of optical photons begin to approach the dimensions of clusters of air molecules. The shorter waves of blue and violet light "see" the clumps of molecules better than do the longer red waves and as a result are scattered much more efficiently. Theory shows that the scattering efficiency of the Earth's atmosphere depends on the inverse of the fourth power of the wavelength. Photons at the short end of human detectability, 4000 Å, are *16 times* as likely to be scattered as those near the long end at 8000 Å. Shortwave solar radiation thus bounces to us from all over the sky, which as a result we see as an intense blue. (Violet light is actually scattered the most, but the Sun's violet component is weak and the eye has little sensitivity to that color.) Conversely, the Sun itself, its blue

Frederick Church's painting Twilight in the Wilderness vividly contrasts the red Sun, its light reflected from the clouds, with the blue sky—complementary effects of the scattering of light by air molecules.

component weakened, is slightly reddened. Solar reddening is most obvious at sunrise and sunset when sunlight must pass through more air to reach observers on Earth: the atmospheric path length to the Sun toward the horizon is 38 times that directly overhead.

Scattering dominates optical interstellar extinction, the grains tending to be bright, with high albedos. We can rather easily determine the manner in which interstellar extinction varies with wavelength, expressed as the "extinction function," by comparing the brightness variation with wavelength of two similar stars, one nearby and unreddened, the other distant and highly reddened. The extinction function depends not on the inverse of the fourth power of wavelength, but—to a rough approximation in the optical—simply on the inverse on the wavelength, λ^{-1}, which implies from scattering theory that the grain diameters are now roughly comparable to the dimension of optical wavelengths. Detailed analysis shows

The extinction function—the dependency of interstellar extinction to wavelength—has a rough λ^{-1} dependency in the optical and goes to zero at long wavelengths (allowing far-infrared and radio waves to penetrate interstellar space with panache). In the ultraviolet, interstellar space becomes remarkably opaque; the 2200-Å bump is caused by amorphous carbon.

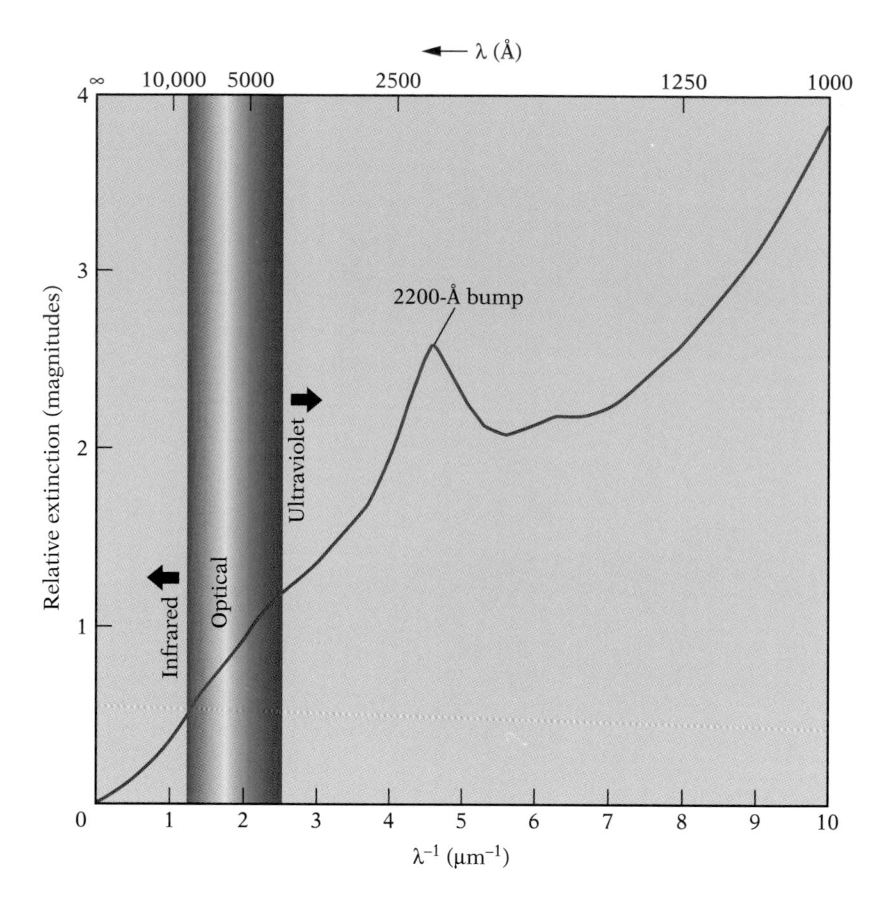

that they have average diameters of a micron (10^{-3} mm) or so (the prescient Russell was correct), with a wide range that takes them from the sizes of large molecules to about 0.3 micron (μm). Unlike the dust on your dining room table, interstellar dust is invisible to the eye.

The spectral intensities of stars and of the emission lines of nebulae, at least of those outside our local region of the Galaxy, are all distorted by the effects of interstellar reddening. For example, the observed ratio of the strength of the 4363 Å line of [O III] emitted by diffuse nebulae to that of the line at 5007 Å always appears smaller than the true ratio. If we do not correct for the effect, we will underestimate the nebula's electron temperature. Once we know the extinction function, we can "deredden" the observations, an initial step in deriving the physical properties and chemical compositions of the nebulae. Nebular brightnesses are not well expressed by magnitudes, which record brightness over a wide range of wavelength that incorporates a varied mixture of nebular emission lines. There is therefore no direct measure of color excess. Instead, the degree of reddening is found by comparison of the observed intensities of the hydrogen lines with those calculated theoretically.

A remarkably low density of these tiny particles produces the observed extinctions: astronomers estimate that, aside from the dense globules and dark clouds, there is an average of only about one grain per cubic meter. Interstellar extinction is important only because the path lengths to stars are so very long. There are 3×10^{16} meters in a parsec, and noticeably reddened stars are hundreds, even thousands, of parsecs away. From the grain density and size we estimate that the solid particles constitute about 1 percent of the mass of the interstellar medium. Interstellar gas and dust seem to be remarkably well mixed, staying at about the same mass ratio except perhaps under extreme conditions.

We would naturally expect the grains—like the stars—to be made of the most common elements. The reddening function and infrared spectroscopy tell their gross compositions. In the ultraviolet, where wavelengths are much shorter than in the optical, the extinction efficiency becomes much larger, greatly dimming ultraviolet starlight and making ultraviolet studies quite difficult. (Astronomers with ultraviolet observing programs using satellite observatories such as the now-retired *International Ultraviolet Explorer* (*IUE*), a flying spectrograph, or the *Hubble Space Telescope* (*HST*), must have extinction estimates for their programmed objects so they know they can be seen at all!) Initial ultraviolet studies of stars showed a sharp enhancement, or "bump," in the ultraviolet extinction function at 2200 Å (0.22 μm). Comparison with laboratory studies showed that the bump is produced by tiny grains only 0.01 μm or so across made of amorphous—that is, noncrystalline—carbon, something akin to ordinary soot or graphite. The bump is useful in that, like color excess, its prominence in the ultraviolet spectra of stars and other celestial objects indicates the degree of total extinction and thus allows accurate dereddenings and estimations of distance.

More detailed chemical compositions are provided by studies in the infrared part of the spectrum. The grains in the line of sight superimpose broad absorption features on the infrared spectra of distant stars. Two of these, at 9.7 μm and 19.0 μm, are related to silicates, grains made of molecules that are compounds of silicon and oxygen. Perhaps most remarkable, we also see an infrared absorption dip that seems to come from diamond dust. Somehow as much as 30 percent of the carbon has been crystallized. The enormous

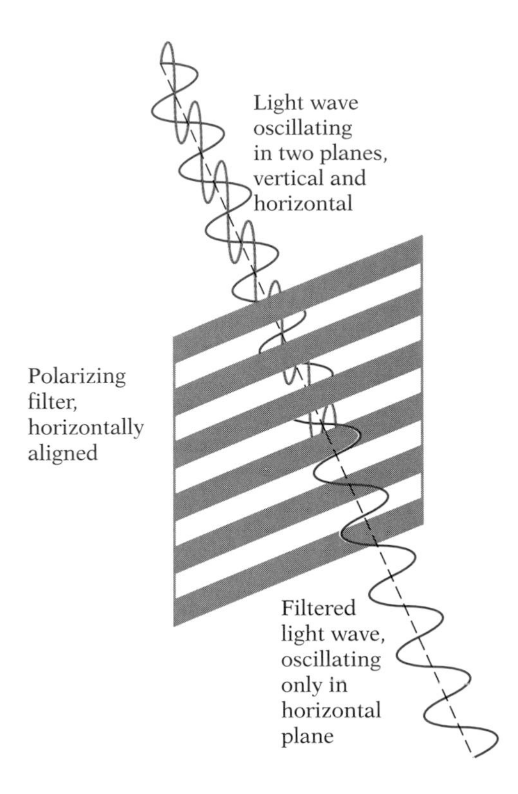

Randomly polarized light passes through a filter that preferentially removes a plane of oscillation, and the light becomes polarized, the waves oscillating together in the same direction.

pressure required most likely comes from shock waves from exploding stars. The diamonds are hardly suitable for interstellar mining, however, as they are of submicron size. The high albedos of the grains also suggest a coating of ice.

In 1949, J. S. Hall and W. A. Hiltner independently discovered yet another property of the interstellar medium: its ability to polarize light. Imagine a beam of light shining toward you. If you could actually see the electromagnetic waves themselves, not just their effects, you would normally observe the myriad photons' waves oscillating in all possible directions perpendicular to the line of sight: up and down, side to side, and all directions in between. However, a variety of natural processes can remove some directions of oscillation, leaving preferred directions and thus "polarizing" the radiation; if the waves oscillate in only one direction, the light is said to be completely polarized.

Reflection at an angle always preferentially polarizes light in the direction parallel to the reflecting surface; that is, the efficiency of reflection is greater for rays oscillating parallel to the surface than for rays oscillating perpendicular to it. If the striking angle is very low, the light is almost completely polarized. A filter in which molecules are preferentially aligned also polarizes light. Drive down a highway with the setting Sun glaring off the road. If you put on polarizing sunglasses in which the polaroid filter is aligned perpendicular to the preferentially reflected plane of oscillation, suddenly the glare is gone and you can see.

The Earth's atmosphere provides another example, as scattering from air molecules also polarizes light. Look at the blue sky with a lens from a pair of polarizing sunglasses. Rotate the lens and you will see that the brightness of the sky varies, showing that its light is partially polarized; the greatest variation is seen in the direction perpendicular to that to the Sun. (Polarizing filters are useful in landscape photography because they can heighten the contrast between sky and scenery.) Cumulus clouds show a similar phenomenon. While wearing polaroids, notice that clouds in directions from which the Sun's rays are scattered to you at right angles appear dark against the sky, whereas those in other directions appear bright.

Starlight is polarized by only a few percent, far below the degree that simple polaroid sunglasses would detect. Its measurement is tricky because reflection by the telescope optics can add unwanted polarization. The degree of polarization crudely correlates with extinction, indicating that interstellar grains are responsible for the phenomenon. Theory shows that to polarize light, the grains must act something like a polarizing filter—that is, they must be elongated and in some way aligned. How can one grain possibly know what another is doing? The most likely answer is that the grains contain metal atoms that are influenced by a magnetic field pervading the interstellar medium. The alignment does not tell us the field strength, but large numbers of observations show that the field runs more or less along the plane of the Galaxy. The rotating Galaxy itself, coupled with the electrical conductivity of its turbulent ionized gases, apparently acts as a gigantic dynamo that generates the field, and the field in turn is revealed by the dust, showing a further intimate relation between gas and grains. Interstellar polarization also tells us more about grain compositions. The silicates, which presumably contain at least some of the metal atoms, may be minerals much like those that make the rocks of Earth—a hint of Earth's link with interstellar material.

SEEING WITH THE INFRARED EYE

If interstellar grains absorb radiation—radiant energy—they must heat up. But, as blackbodies, they must also radiate, their temperatures established at the point at which the heating rate equals the cooling rate. The final temperature of a grain in equilibrium will thus depend on the intensity of radiant energy, on the grain's radiating area, and on the chemical composition. Grains in the nebulae are far from their illuminating stars, sometimes parsecs away, and the intensity of radiation is low. The heating rate in interstellar space, even within reflection nebulae, is therefore also low, so the grains are generally cold, with temperatures far below those needed to make them glow in the optical.

Cold as they are, however, the grains are warm enough to radiate in the infrared, and telescopes with infrared detectors are able to observe their glow—but with only partial success, because the Earth's atmosphere imposes a severe problem. Except for a few transparent bands, the air is generally rather opaque to infrared radiation, because of absorption by the greenhouse gases—carbon dioxide and water vapor—that keep us warm. As a result, infrared astronomers have established high-altitude observatories sited above as much air as possible and have even outfitted high-flying aircraft with telescopes; until its retirement, the *Kuiper Airborne Observatory* flew aboard a C141 Starlifter, and a successor to be built into a Boeing 747 is in development.

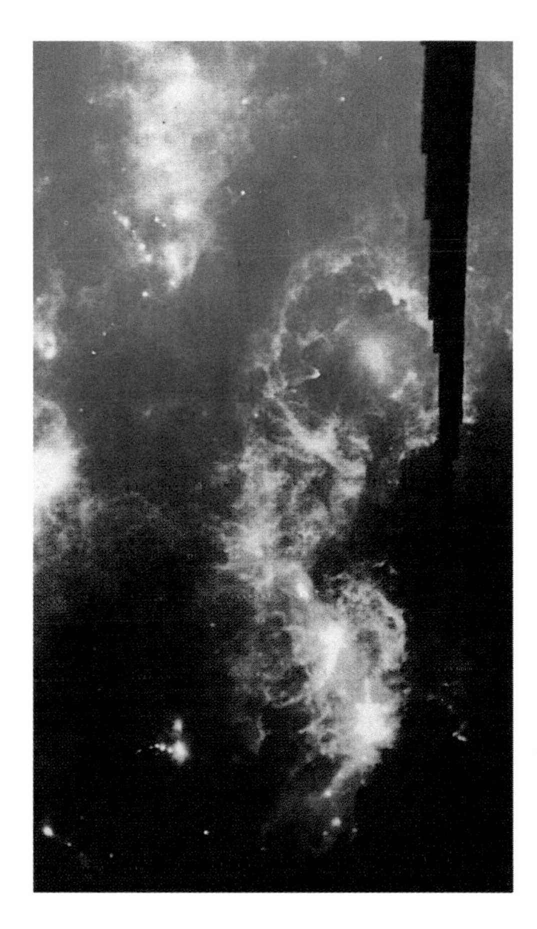

The Infrared Astronomical Satellite viewed an Orion filled with cool, glowing dust. The Orion Nebula lies within the bright arc near the bottom; the dust that forms the Horsehead Nebula makes the arc above it. A huge ring near the top surrounds the bright O star Lambda Orionis. In the background is wispy "infrared cirrus."

For best results, however, telescopes are flown in satellites. The pioneer was the *Infrared Astronomical Satellite* (*IRAS*), launched in 1983. Cooled by liquid helium to only 3 K (any warmer and radiation from the telescope would have overwhelmed that from the celestial objects being observed), *IRAS* mapped the entire sky at wavelengths of 12 μm, 25 μm, 60 μm, and 100 μm. These wavelengths correspond to peak emission for blackbodies at temperatures of 240 K, 115 K, 50 K, and 30 K respectively, appropriate to the cold conditions of interstellar space. By the time the helium was exhausted, a year after launch, enough data had been gathered to maintain an institute (the Image Processing and Analysis Center at Caltech) that in the late 1990s is still processing information, still helping astronomers make discoveries in the frigid infrared sky.

The *IRAS* view, completely unrecognizable to the eye, allows us to see the glow of red stars, of radiation from diffuse nebulae, and of the ubiquitous heated dust. Instead of seeing the dust as dark features against a bright sky, *IRAS* saw it as bright, the intensity depending on temperature. The temperatures of the radiating sources are easily found by comparing the ratios of the intensities (I) at the different wavelengths—for example I(60 μm)/I(100 μm)—against those expected from blackbodies. The most surprising dusty features were not the globules and rifts—which are internally so cold that they radiate very little even in the infrared because the dust screens itself from heat-producing stellar radiation—but infrared cirrus, looking for all the world like the wispy "horsetail" clouds seen in a blue summer sky. From the 60 μm/100 μm intensity ratio, the temperature of the cirrus is only 20 to 30 K—cold indeed.

Remarkably, however, the 12 μm/25 μm intensity ratio showed a very different picture. At these wavelengths we see grains that are heated to hundreds of degrees Kelvin, warmer than the Earth (where the average temperature is 288 K). There is not enough stellar ultraviolet energy to maintain equilibrium at such a high temperature, and the grains' heat is taken as further evidence that they are very small, of the size that produce the 2200 Å extinction bump. They are so small that they heat suddenly upon being struck by individual stellar photons, spiking to high temperatures from which they immediately cool. These results demonstrate the existence of a remarkable variety of grains as well as the complexity of the interstellar medium.

But the infrared astronomers (who are themselves quite visible in the optical) are hardly content to rest on their successes. The infrared gives us a profound probe into space and the next few years should see important results from the recently launched *Infrared Space Observatory* (*ISO*) and, we hope, from the long-planned *Space Infrared Telescope Facility* (*SIRTF*), one of NASA's series of "Great Observatories." We will undoubtedly see things we never dreamed were even there—extending the experience, almost a century ago, of the German astronomer Johannes Hartmann, whose work continues our story.

4

Opening the Spectrum

An association of hot stars in the galactic disk (marked by the Xs) blows a "chimney" in the interstellar medium through which gas pushes into the galactic halo.

We learn in school that science works by the "scientific method," that experiment or observation leads to theory, theory predicts phenomena that can be experimentally tested, experiments lead to improvements in theory. However, scientists are equally at home in the country of Serendip, whose three princes set out to look for one thing and found another. Herschel thought he could detect parallaxes by examining close pairs of stars; instead, he found double, or binary, stars, pairs whose members orbit each other, and thus opened a window onto stellar and—more surprisingly—*interstellar* astronomy.

Interstellar Ghosts

The components of classic "visual" binaries are angularly far enough apart to be resolved by optical telescopes. Such wide angular separations mean large physical separations, so the orbital periods are correspondingly long: years, decades, centuries, even millennia. Commonly, however, binary components are so close that they cannot be visually separated, appearing as one through the telescope. Since (from Kepler) the square of an orbital period is proportional to the cube of the orbital dimension, the periods of these stars are short—weeks, days, even hours. Each star produces its own spectrum, and the two combine into one. If the orbital plane is tilted to any degree into the line of sight, and if the stars have comparable brightnesses, the high velocities allow us to watch each individual spectrum oscillate back and forth in wavelength as a result of periodic Doppler shifts. Even if one component overwhelms the other, we know the star is a binary because the spectrum of the brighter star wobbles to and fro.

Delta Orionis, the right-hand jewel of the hunter's belt, an example of such a "spectroscopic binary," consists of a B0 giant and an O9 dwarf, each of whose absorption lines continually shifts in wavelength with a 5.7-day period. However, Johannes Hartmann, when investigating the pair in 1904, discovered a narrow absorption line of ionized calcium whose wavelength was fixed. Because the line was stationary, it could not have been produced in the atmospheres of the orbiting stars. "It points," he wrote, "to the presence of an absorbing layer of gas not in immediate connection with the star," and he concluded that "it is not unlikely that the cloud stands in some relation to the extensive nebulous [bright] masses shown by Barnard to be present in the neighborhood." In a long tradition of serendipitous astronomical discovery, Hartmann—though he did not

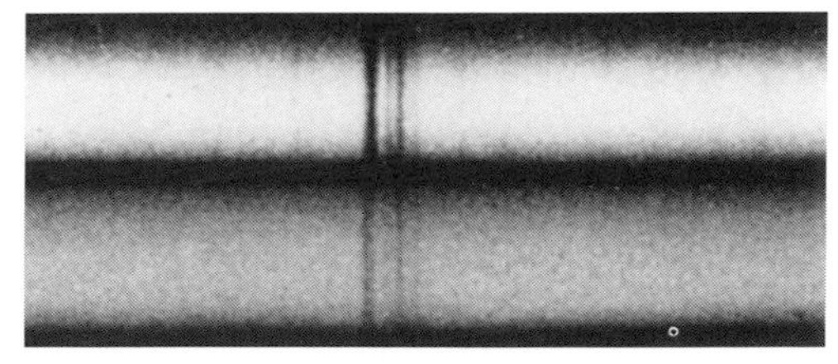

Multiple interstellar lines of ionized calcium and neutral sodium are observed against the spectrum of ϵ Orionis, the middle star of Orion's belt.

know it—had discovered the gaseous interstellar medium, that *not* confined to diffuse nebulae.

Calcium is not a terribly abundant element, comprising only about 10^{-8} the number of atoms in the Sun. The interstellar lines of Ca^{+}, at 3934 Å and 3968 Å (just at the edge of human vision), are noticeable only because they arise directly from the ground state and because all Ca^{+} ions are capable of absorbing at these wavelengths. Vesto Slipher realized that if the ground state "resonance lines"—the lines produced by an atom or ion that arise between the two lowest energy levels—of Ca^{+} are visible, the similar resonance lines of neutral sodium (in the yellow-orange at 5890 Å and 5896 Å) should be visible as well. In 1919, these lines too were found in the spectrum of δ Orionis.

Similar absorptions were subsequently found in the spectra of single (nonbinary) hot stars, where they were recognizable as interstellar interlopers because their line-of-sight, or radial, velocities, determined from their own Doppler shifts, were not the same as the stars' radial velocities. The interstellar lines are difficult to see in the spectra of cooler stars because the complexity of stellar spectra increases as temperature drops. The intrinsic spectra of the relatively rare hot O and B stars, however, are supremely simple, thus providing a fertile ground for interstellar study. By the mid-1920s it was becoming clear, as a result of the numbers and natures of the narrow lines, that they were not caused simply by a medium in the neighborhoods of the stars but were indicative of a general interstellar gas. Over the next 40 years astronomers added to their inventory the red resonance lines of neutral potassium (at 7665 Å and 7699 Å) and neutral calcium (4226 Å), as well as lines of neutral iron and ionized titanium. Portending great things to come, they even found interstellar absorption lines produced by the simple molecules CH, CH^+, and CN (cyanogen).

Of equal importance to the lines themselves were the discovery in 1936 that many of them were doubled and then observations of very high spectral resolution showing that most were *multiple,* each component having a slightly different wavelength, each Doppler-shifted by a different constant velocity. The absorbing medium is therefore not uniform but organized in independently moving ghostly gaseous *clouds.* They were not the globules, the dark clouds of the last chapter—there was nothing to be seen hiding or obviously dimming the stars. Analysis of hundreds of stellar spectra shows that, like most of the interstellar medium, these clouds are confined to the galactic plane and suggests there are, roughly, five or so per kiloparsec, spread very irregularly, with typical dimensions of a few parsecs.

However, we can derive only limited information about the clouds from spectra showing what amount to trace elements. As in diffuse nebulae, hydrogen should be by far the most abundant element. Unfortunately, we cannot see hydrogen in absorption against optical stellar spectra because the temperatures in the clouds are too low to raise the electrons by collision into the second orbit or energy level, from which the visible Balmer series arises. Moreover, only clouds seen against bright O and B stars can be so examined. We need to expand our spatial and spectral views.

Radio to the Rescue

The "orbital" structure of hydrogen described earlier is vastly oversimplified. In the "classical" view of the atom (the Bohr model, which is not really accurate, but an understandable schema), each energy level except the first can be split into sublevels, into a "fine structure," in which the electron orbits can be elliptical and can take on different specific, quantized values of eccentricity. Counting outward, for each energy level (or orbit) *n*, there are *n* of these sublevels, each differing very slightly in energy. Only transitions between neighboring sublevels are allowed. All the hydrogen lines except those of the Lyman series of transitions (which connects to the first level, or ground state) are therefore multiple. But because the fine-structure levels are so close together, the resulting fine structure of a Balmer (or other hydrogen) line cannot ordinarily be seen. This kind of structure is part of the reason for the complexity of the spectra of atoms heavier than hydrogen: the interactions among the increasing number of electrons in a sense pull the sublevels apart, producing a multiplicity of lines.

Electrons and protons have a property called spin. Like orbital energies, spin can take on only certain values—that is, it is quantized. The movement of electric charges produces magnetic fields. Because protons and electrons carry such charges, the spins create tiny fields, as does the electron's orbital motion. The proton's spin-generated magnetic field interacts with the electron's spin-orbit field. There are two possibilities. The electron in a hydrogen atom can spin in the same direction as its proton, axes aligned, or "parallel," or it can spin in the opposite direction, "antiparallel." If the spins are parallel, the magnetic energies add together and the total energy of the atom is slightly higher than if the spins are antiparallel. Thus the ground state (and other levels as well) is split into a pair of "hyperfine" levels.

The ground-state hyperfine splitting is very small. The energy required to ionize a hydrogen atom is 13.6 electron volts (a 100-watt light bulb uses energy at the rate of 1.6×10^{21} eV per second). A photon with this energy has a wavelength of 912 Å, the Lyman limit. Ultraviolet photons with shorter wavelengths (higher energies) ionize hydrogen and thus are responsible for the glow of diffuse nebulae. The upper hyperfine level of the ground state is a mere 5.9×10^{-6} eV above the lower level, an energy that corresponds to a radio photon with a wavelength of 21.049 cm.

In 1944, the Dutch astronomer Hendrick van de Hulst predicted that this 21-cm emission line from neutral interstellar hydrogen might be detectable. As you read, the atoms in the room are colliding

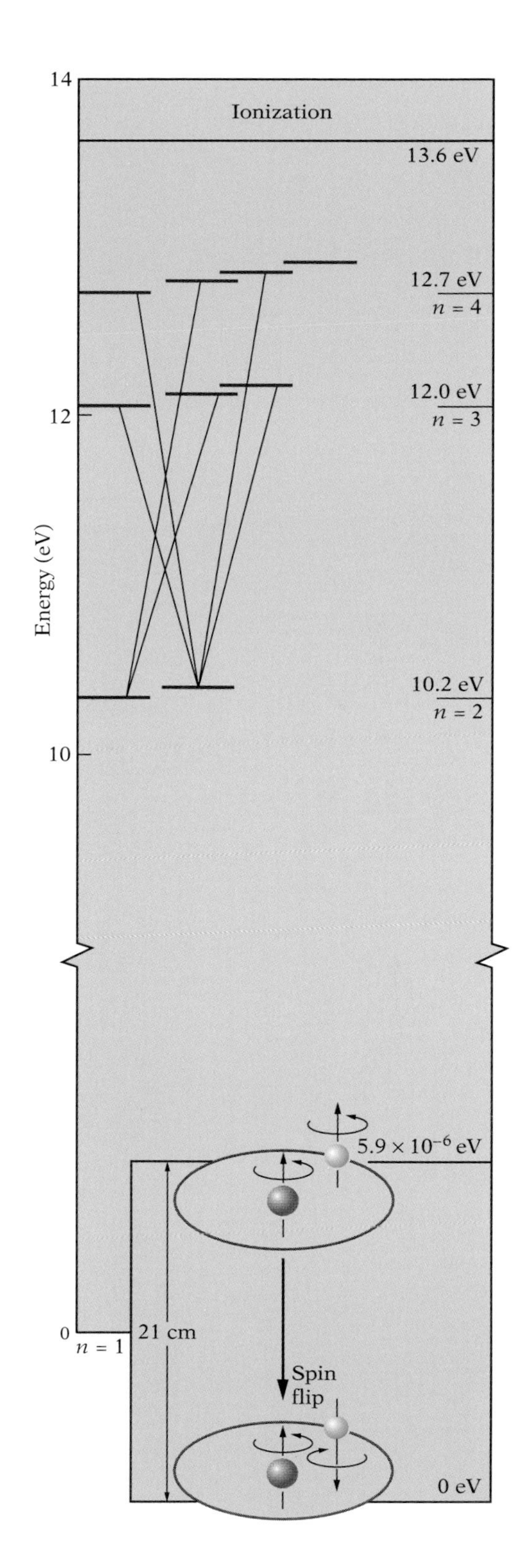

Upper hydrogen energy levels are split by differing orbital eccentricities. Level 1 is split by magnetic forces: in the upper state, the electron and proton spin similarly; in the lower, oppositely. The transition between the two produces the 21-cm line.

at a fierce rate, a given atom hitting another roughly every billionth of a second. But as temperature goes down the atoms move more slowly, and as the density decreases there are fewer opportunities for interaction, so the collision rate drops. In the cold thin gas of dusty interstellar space, where densities are in the hundreds, or even tens, of atoms per cubic centimeter (as opposed to 10^{19} cm^{-3} in your room), a lonely hydrogen atom will typically wait for hundreds of years before encountering another.

An atom has nothing if not time, however, and collide it will. If the hydrogen atom is in the bottom hyperfine state, its proton and electron spins antiparallel, the collision can reverse the spin, sending the electron upward into the parallel state. There are two possibilities: the electron can jump downward, reversing spin spontaneously, radiating the energy as a 21-cm photon, or it can wait a few hundred years for another collision that will knock it back down. The downward radiative transition, however, is "forbidden." An electron in energy level 2 (the level that relates to the Balmer lines) will jump to level 1 in 10^{-8} s; one in the parallel hyperfine state will typically wait *11 million years* to make the flip to antiparallel. Collisions, obviously, dominate, and because of the quantum statistics of the atom, there will be three times as many electrons in the upper hyperfine state as in the lower. Yet within the hundreds-of-years wait for the downward collision to occur, there is about a 1:20,000 chance that the radiative transition will take place first (state lottery odds are vastly worse). And there is so much hydrogen in interstellar space that the 21-cm line should build up great strength: even at a density of 1 atom cm^{-3}, a square centimeter column a kiloparsec long will contain over 3×10^{21} hydrogen atoms!

The 21-cm line was found by Harold Ewan and Edward Purcell in 1951. It is by far the strongest interstellar emission line, easily powerful enough now to be picked up by a knowledgeable radio amateur. Observation and analysis of the line revolutionized the study of the interstellar medium and of the Galaxy itself. No longer were we limited to what we could learn from the spectra of metals seen only by luck against the spectral backgrounds of rare stars or by emission by ionized gas; now we could find, and examine, the total distribution of neutral hydrogen.

The observed 21-cm emission line has a complex shape with many peaks, showing that it consists of the overlapping emissions from many different clouds and regions of gas moving at different radial velocities. They would logically seem to be the same clouds that produce the interstellar absorption lines of calcium and sodium. Occasionally a cloud will be positioned in front of a bright distant radio source; in this case the foreground hydrogen *absorbs* the 21-cm line, as it absorbs the optical atomic lines. The interstellar structures that produce the absorptions and emissions of neutral hydrogen are commonly known as H I regions, in parallel with the common practice of calling ionized diffuse nebulae H II regions.

The strength of a line—the radiant energy in an emission line or the amount of radiation absorbed from the background continuum by an absorption line—depends on two things: the number of atoms that lie along the line of sight (the column density, the number of atoms in a tube with a cross-sectional area of 1 square cm) and on the probability, calculated from theory or measured in the laboratory, that an individual atom will emit or absorb the relevant photon. The problem is complicated by the fact that an emission cloud will absorb some of its own photons and that any absorption line will be partially filled in by emitted photons. In addition, once an absorbing cloud picks up all the incoming photons, it obviously cannot absorb any more and is thus fully black, or saturated. Such clouds, in which the radiation, of whatever wavelength domain, is significantly re-absorbed or re-emitted, are said to be optically thick.

The interplay between absorption and emission allows the determination of conditions within the clouds. Complex mathematical

A galactic map of H I at a radial velocity of 0 km/s from 21-cm observations shows not so much clouds as a complex nest of interconnecting sheets, bubbles, and filaments.

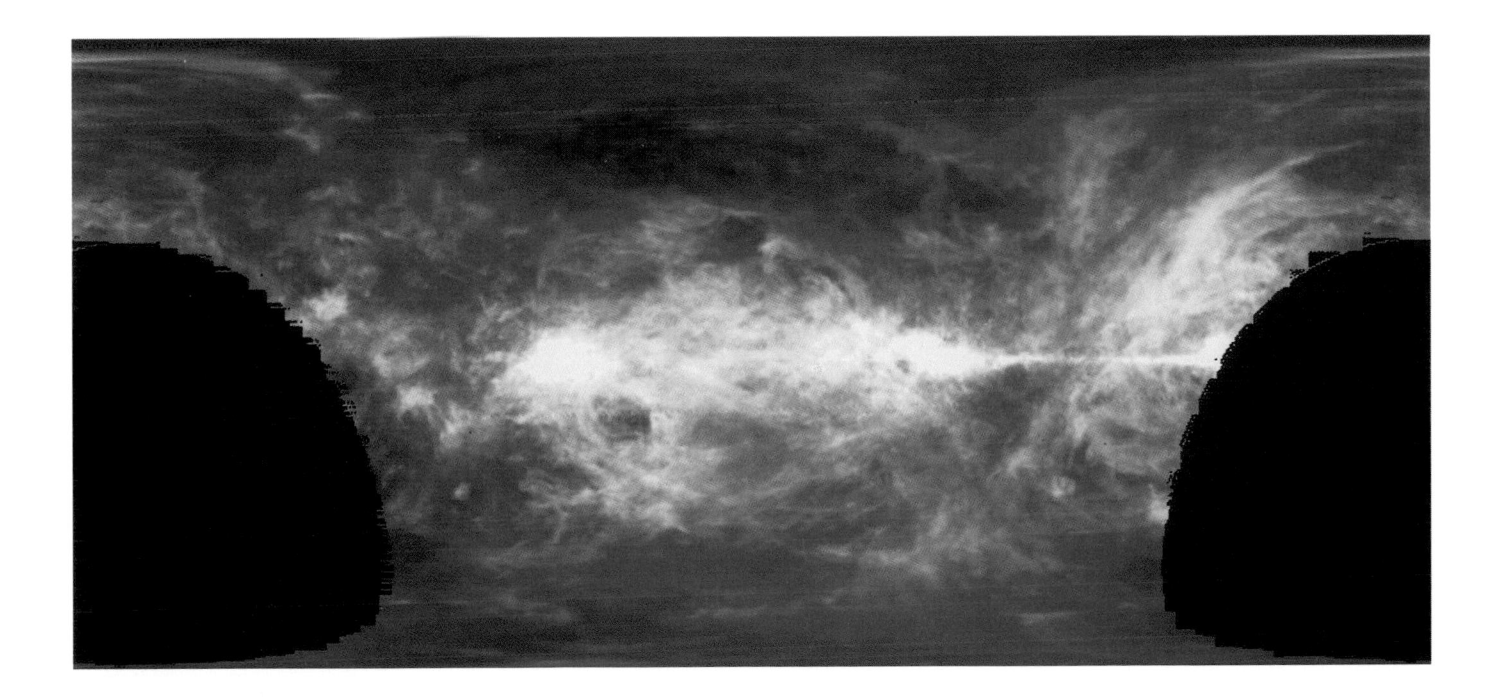

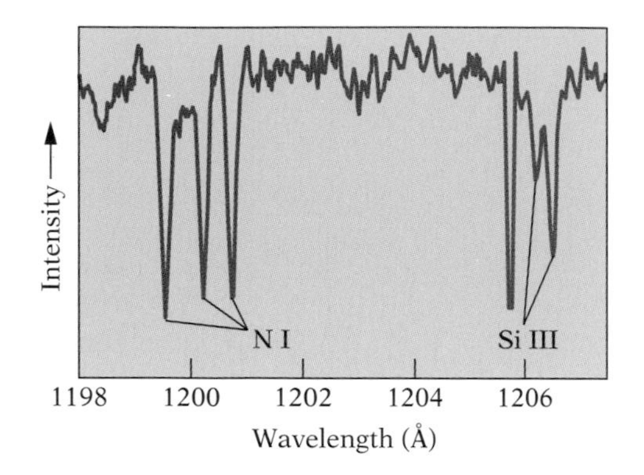

A spectrum taken with the IUE satellite shows a variety of interstellar absorptions in the ultraviolet.

analyses of the relative strengths of the absorption and emission lines produced by neutral hydrogen clouds reveal temperatures averaging about 100 K, with a range from 40 K or so to about 200 K. The H I clouds are cold but not *that* cold, and as a result they are considered to be the "cool" component of the interstellar medium. From the column densities and an estimate of the clouds' dimensions of one to a few parsecs, we find densities of a few tens to a few hundreds of atoms per cubic centimeter, generally below that of the classic diffuse nebulae. Cloud masses (calculated from the known mass of the proton) are therefore typically in the range of a few tens to maybe a hundred times that of the Sun.

Critical to a more complete understanding of the clouds is knowledge of their chemical compositions, for which we return to the absorption lines produced against stellar backgrounds by elements other than hydrogen. The optical spectrum provides us with valuable interstellar nuggets, but the mother lode is in the ultraviolet. First to mine it was *Copernicus,* the satellite that in 1972 carried Princeton University's ultraviolet spectrograph. It was followed in 1975 by the fabulously successful *International Ultraviolet Explorer,* and then by the *Hubble Space Telescope,* launched in 1990 and successfully repaired in 1993. It is in the UV that we see lines of zinc, magnesium, silicon, manganese, and a bevy of others.

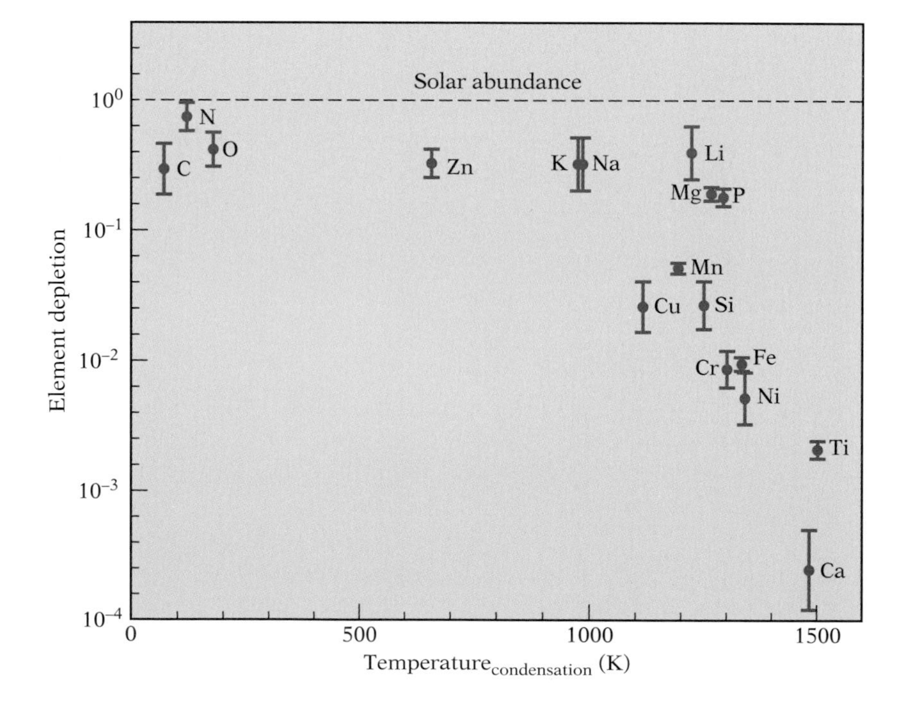

The depletions of interstellar elements relative to solar abundances correlate with condensation temperature, strongly suggesting that the depletions are caused by solidification onto grains from the gaseous phase.

Measurements of the column densities combined with those of hydrogen allow the determination of the ratio within the clouds of a given element to hydrogen. Comparison of the results with the solar composition reveals a decidedly odd pattern in which almost all the observed elements heavier than hydrogen are depleted—that is, their relative abundances are lower than in the Sun. The abundance of nitrogen (that is, the N/H ratio) is just about "normal," that is, similar to that found in the Sun (and in the Earth and meteorites as well). But oxygen and carbon are low by a factor of 2 or 3, possibly partly because of composition inhomogeneities within the Galaxy. The depletions for some elements, however, are extraordinary: iron is down by a factor of 100, and titanium and calcium by factors of 1000. Moreover, the denser the cloud, the greater the level of depletion.

Where are the missing atoms? Since we know that the Sun and Earth are made of interstellar matter, the atoms have to be somewhere. The critical clue is that the depletions correlate with condensation temperature, the temperature at which an element in the gaseous state can solidify. Refractory elements, like calcium, condense at high temperatures—that is, they form solids easily under energetic conditions; on the other hand, volatile elements, like nitrogen, require *low* temperatures before they can condense. At the temperatures of the H I clouds, the refractory elements are producing solids. Interstellar grains contain metal atoms that help the grains to be aligned by the galactic magnetic field, and as a result polarize starlight. The thief in this interstellar robbery must therefore be interstellar dust! Composed principally of carbon and silicates, the grains must be subject to a complex chemistry that draws heavy atoms out of the gas. By examining the gas, we have discovered more about the composition of the dust, more firmly than ever linking dust and gas in an inextricable partnership.

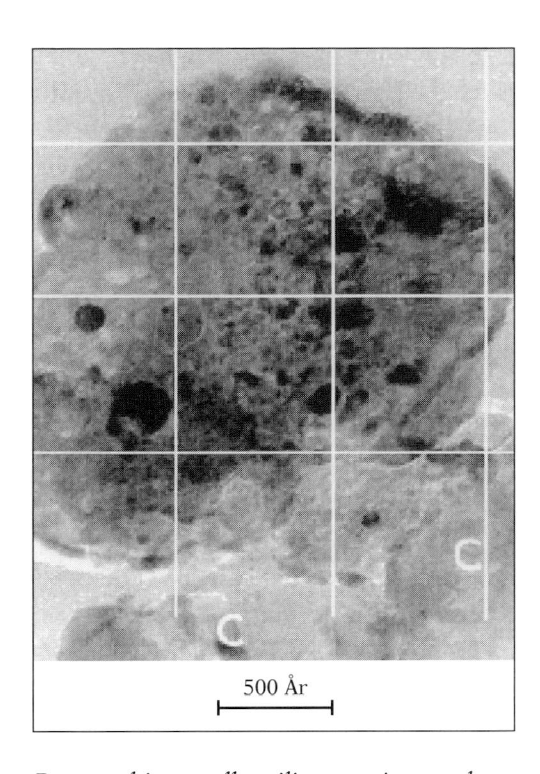

Preserved interstellar silicate grains can be found inside carbonaceous interplanetary dust particles (the part labeled "C"). Within the grains are dark metallic inclusions that are apparently responsible for aligning the grains via the Galaxy's magnetic field and for producing interstellar polarization.

The Big Picture

We cannot comprehend the clouds, their origins and structures, without knowing their place within the Galaxy. The very existence of the band of light we call the Milky Way shows us that most of the Galaxy's stars are distributed in a disk. The clouds—neutral H I clouds, dark clouds, diffuse nebulae, reflection nebulae—are also concentrated within this galactic plane. Our knowledge of the structure of this disk, and the way in which the clouds are set into it, comes in large part from analysis of the clouds themselves.

In 1794, William Herschel began to examine the structure of the Galaxy. He made the assumption—naïve, as we know now—that all

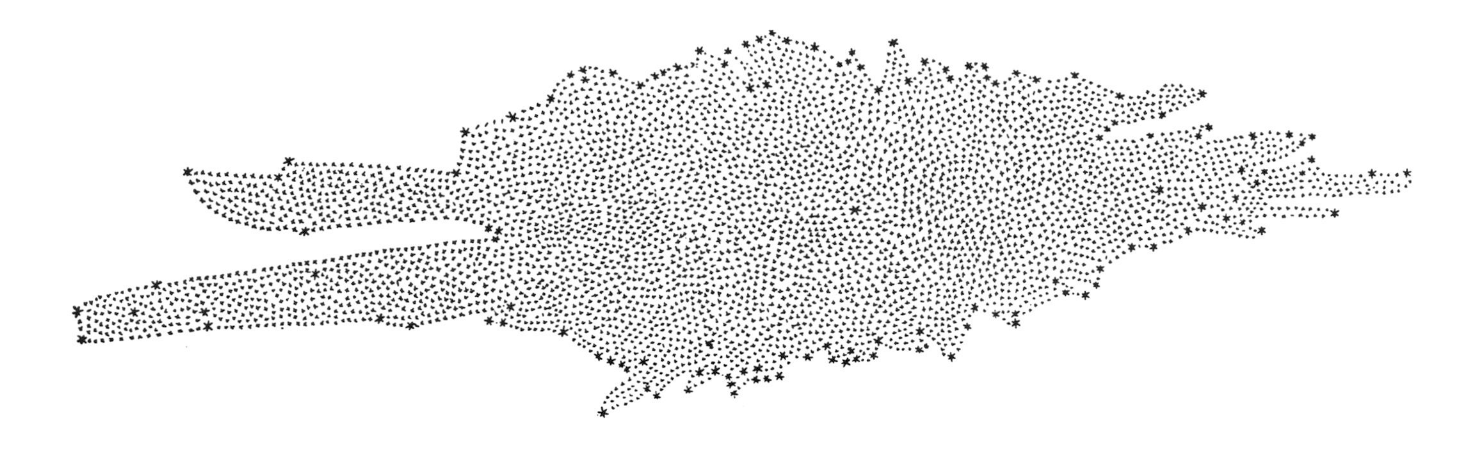

The Galaxy as mapped by William Herschel was a flattened structure with the Sun near the center. The effects of interstellar extinction were unknown to him.

stars have the same intrinsic brightness, and therefore their apparent magnitudes should reflect their distances. By counting the number of stars within each magnitude division in different directions, he produced a map. His model resembled a grindstone, a thick disk, with the Sun near the center.

Herschel was right about the Galaxy's flattened nature. However, he was entirely wrong (as were generations of astronomers to come) about the solar location. Interstellar extinction, a phenomenon unknown to him, dooms any optical study of large-scale galactic structure that is confined to the disk. Even with an average local extinction within the galactic disk of only about a magnitude per kiloparsec, our optical view toward the Galaxy's interior is restricted to only a few kiloparsecs, as closer to the galactic center the extinction increases enormously—the galactic nucleus is hidden behind 30 magnitudes of it (one photon in a trillion gets through). As a result, we effectively see only the disk's local neighborhood on our side of the Galaxy—an important view, but nowhere near broad enough to establish how the Galaxy is put together.

About 1918, Harlow Shapley constructed a picture of the Galaxy by using its globular clusters. Since the globular clusters are in the galactic halo, many at high angles from the galactic plane where only a little dust intervenes, they can be seen most of the way across the Galaxy. He estimated their distances from variable stars (those whose brightnesses change) within the clusters—the variation periods yield the stars' intrinsic luminosities. Shapley made the then bold assumption that the centroid of the distribution of the clusters was identical to that of the stars in the disk. He originally found the Sun to be 13 kpc from the center, but his calibrations of the distance-yielding variable stars were faulty and he had not allowed for the dimming effects of the dust. Modern analyses, which take into account the studies of many other kinds of halo objects, place the Sun about 8 kpc from the center.

Radio waves, which penetrate the haze of interstellar dust perfectly, and specifically the 21-cm line, which provides a wavelength and (through the Doppler effect) a velocity reference, allow us to navigate within the galactic disk. We first establish a coordinate system with galactic latitude and longitude based on the galactic equator, the midline of the Milky Way. The center, at 0° longitude, is taken to be a bright radio source in Sagittarius known as Sagittarius A, which because of its brilliance was long ago identified as the exact center of our Galaxy. (Most astronomers now believe that its core, called Sagittarius A*, is a massive black hole, the radiation coming from compressed and heated matter falling into it.) Galactic longitude is measured along the equator from Sagittarius A northward toward Cygnus; galactic latitude is measured perpendicular to the galactic equator toward the north and south galactic poles that lie in the constellations Coma Berenices and Sculptor respectively.

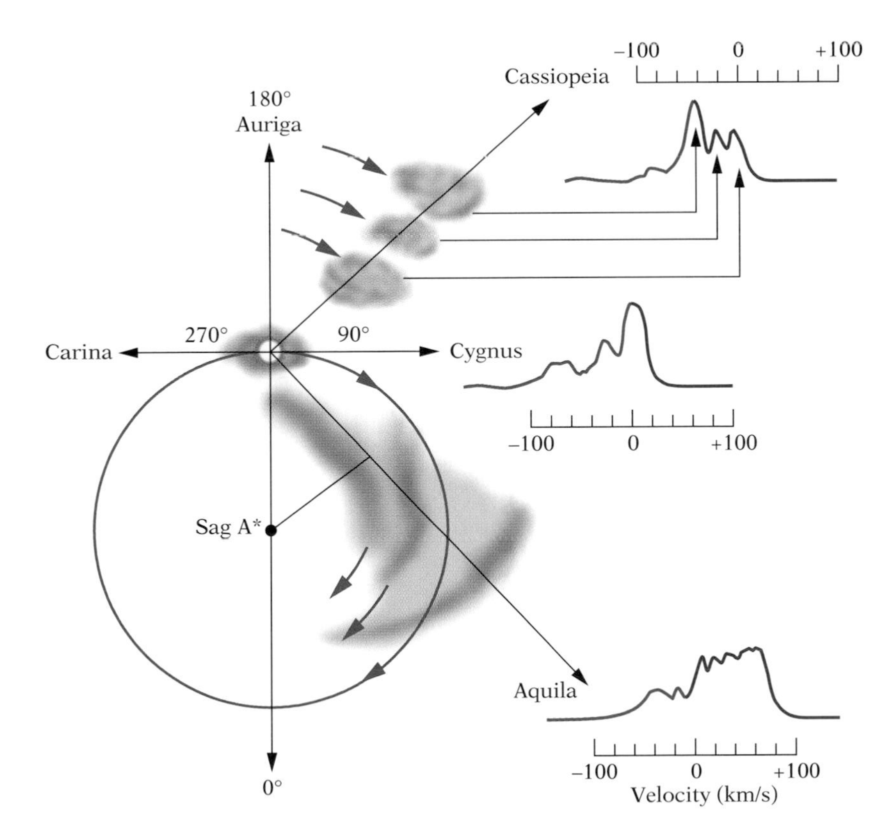

Galactic longitude is measured from the galactic center in Sagittarius counterclockwise. A radio telescope looks into the galactic disk in three directions to view complex 21-cm lines (expressed here in velocity units instead of wavelength) caused by a superposition of clouds moving at different speeds. Toward the inner Galaxy, in the sample direction toward Aquila, the maximum radial velocity relative to the Sun occurs closest to the galactic center, a distance easily found from the right triangle. Correction for the velocity of the Sun gives the rotation speed at that distance.

The Sun, and the other nearby stars, orbit the Galaxy under the influence of the combined gravity of the stars interior to their paths. To find how our Sun moves, we look toward objects—distant star clusters and nearby galaxies—that do not participate in rotation around our galactic center and have random movements relative to us. Their average radial velocities should therefore reflect the orbital velocity of the Sun around the galactic center. Think of walking through a crowd of people at a party, each having different and random directions of motion. Those moving in the direction of *your* motion have the maximum average approach velocity, those moving toward your rear the maximum recession velocity. The maximum average velocity of approach to the Sun, 240 km/s, is in the direction of Cygnus, near longitude 90°; the maximum velocity of recession is consistently near 270°. The Sun therefore appears to be moving at some 240 km/s around the galactic center. Observation of the motion of the Sun relative to the local stars allows us to learn the eccentricities of the local orbits and thus that the velocity of a circular orbit at the solar distance from the galactic center would be about 220 km/s.

We now have one point on the galactic rotation curve, the graph of rotation velocity plotted against distance from the galactic center. Next point your radio telescope, the receiver tuned to 21 cm, into the disk, galactic latitude 0°. Pick a longitude: if you select, say, 35°, you peer into the heart of the Great Rift in Aquila. The resulting spectrum line is a complex blend of the radiations from all the clouds along the line of sight, which are assumed to be moving on circular paths. The radial velocities you observe will be the differences between the projections of the rotational velocities onto the line of sight and the similar projection of the solar velocity. The maximum radial velocity will occur for those clouds all of whose orbital motions are along the line of sight, the ones that lie on the tangent to a circle around, and closest to, the galactic center. The lines Sun–cloud, Sun–center, cloud–center make a right triangle. Since we know the Sun–center distance, we can find the cloud–center distance from the galactic longitude and simple trigonometry (in our example, 4.6 kpc). We have measured the radial velocity of that point only relative to the Sun; addition of the Sun's projected velocity onto the line of sight yields another point on the rotation curve.

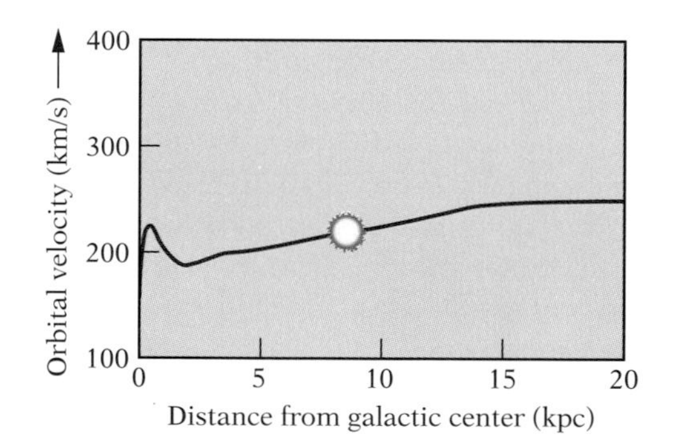

The observed rotation speed of the Galaxy is fairly constant with distance from the galactic center.

A sweep through the galactic plane within the semicircle facing the galactic center, from longitudes 270° through 0° at the center to longitude 90°, allows the establishment of the entire rotation curve interior to the solar orbit. Obviously, this tangent method will not work outside the solar orbit. But there the dust is thinner and we

The nearby galaxy M 100 shows off beautiful spiral structure similar to that found in our own Galaxy.

can see stars associated with various interstellar clouds that yield their distances through the stellar spectra by the method of spectroscopic distances, allowing the rotation curve to be extended outward to over 20 kpc from the center. Within rather broad errors, the rotation speed stays relatively constant at about 200 to 250 km/s from roughly 3 kpc from the center as far out as we can see.

The return in knowledge from this relatively simple but time-consuming observational exercise is profound. When we plot the positions of local young stars—the hot O and B stars we can see, those within a few kiloparsecs—we find they are laid out in broad streams that are nearby sections of spiral arms of the kinds we see in other galaxies. The 21-cm radiation, which arises from sources farther away, reveals spiral arms across the whole Galaxy. As long as the orbits are circular, the radial velocities of the clouds along any line of sight in the galactic plane depend strictly on distance from Earth, and consequently on distance from the Galaxy's center. Once

The H I clouds in the Galaxy trace out thin spiral arms.

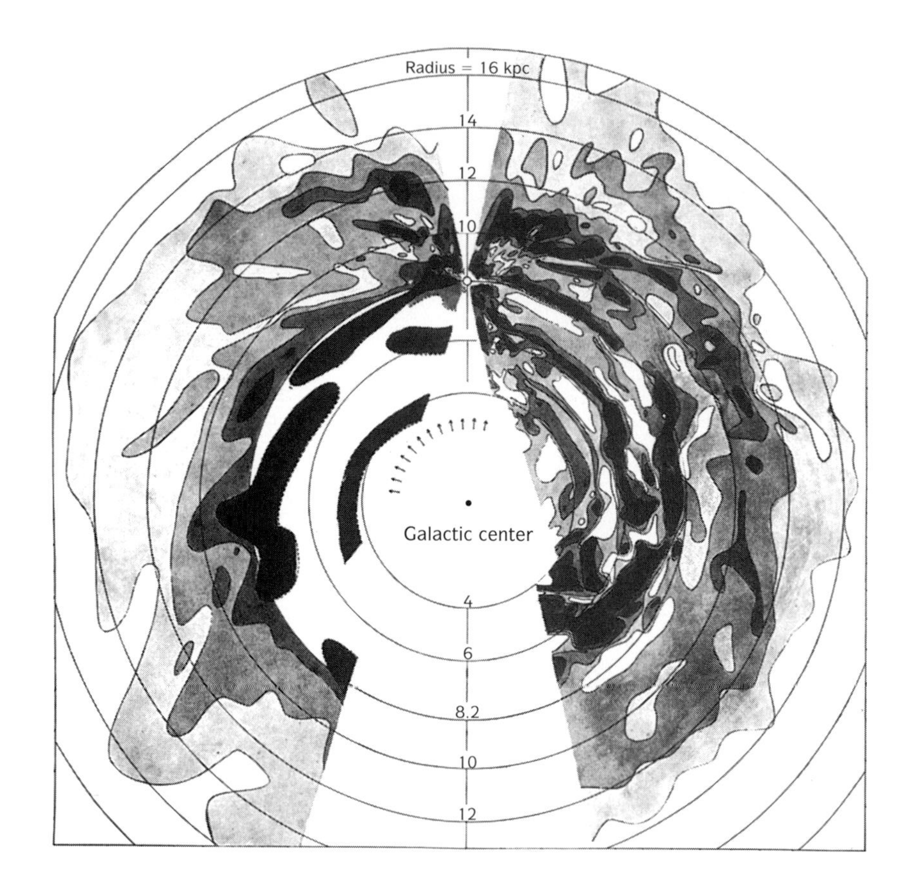

we know the rotation curve, we can locate the clouds and plot their positions, even outside the solar orbit. The result is a map of thin neutral-hydrogen spiral arms that dramatically extends the optical picture. That young O and B stars lie in spiral arms shows that the arms are their birthplaces; that interstellar clouds also define the spiral arms provides a further link between the clouds and star formation.

The arms trail in the direction of galactic rotation, as if they are winding up, and cannot be permanent features. Though the theory of spiral arms has proved difficult and is not yet complete, the arms appear to be density waves propagating through the system. A disturbance within a disk galaxy, caused either from within or without, produces a clumping of matter whose gravity produces additional clumping that moves through the mass, the galaxy's rotation drawing the waves into a spiral form. The matter in the arms comes and goes as the galaxy rotates, the arms continually changing their pattern as well.

The rotation curve also allows a measure of the Galaxy's mass. Newton showed that any gravitating body with a spherical distribution of matter acts as if all its mass were concentrated at the center. If the Earth's mass were balled into its very core and you stood on a platform 6500 km high (the Earth's radius), you would weigh the same as you do now. The solar orbit responds to the mass of the Galaxy that lies interior to it. While the Galaxy is hardly spherically symmetric, to a simple approximation the Sun and the inner Galaxy make a two-body system, the Sun behaving as if the Galaxy's mass within 8 kpc were placed in a ball at the center. We can therefore apply Kepler's laws of planetary motion as modified by Newton to determine roughly the mass of the inner Galaxy, as we do for binary stars.

The Sun orbits the galactic center in a rough circle with a radius of 8 kpc and a velocity (reduced to reflect circularity) of 220 km/s, resulting in a period of 225 million years. Conversion of units to seconds and meters yields a mass internal to the solar orbit of somewhat under 2×10^{41} kg, or about 90 billion solar masses. Assuming that most of this mass is in stars, we find from star counts that we need add only about 20 percent more to obtain the total stellar mass of the entire Galaxy. Most of the stars in the Galaxy are low-mass dwarfs, the average about half a solar mass. A galactic mass of 100 billion solar masses therefore implies the existence of some 200 billion stars. Radio and optical observations show that about 10 percent of this mass is in the gas and dust of interstellar matter, a plentiful supply for the formation of new stars.

Between the Clouds

So far in our exploration, the interstellar medium seems to consist largely of clouds of gas and dust more or less confined to spiral arms. If a cloud is dense it may appear to us as some form of dark nebula; if near a star, as a reflection nebula; if near or surrounding a sufficiently hot star, as a diffuse nebula; if of lower density, as an emitter of 21-cm radiation and an absorber of interstellar lines. What keeps all these clouds together? Why are there such things as clouds? Are they confined and relatively permanent, or are they expanding and dissipating? If confined, what is the confining force? If dissipating, why are there clouds at all? Do they merge, break up, and re-form? If so, why? Such questions force us to probe deeper, to examine not just the inter*stellar* medium but the inter*cloud* medium, presaged by Trumpler's discovery of general interstellar absorption.

The existence of so many clouds and the vaguely spherical shapes of so many of the globules suggest some kind of permanence. Any gaseous body exerts an outward push, a pressure (P) that is proportional to the product of the density (expressed by the number of particles per unit volume, N) and the temperature (T), or $P = N\mathrm{k}T$, where k is a constant. Stability requires an inward force to counter the outward push. For a star the inward pull of gravity exactly counteracts the outward push of gas pressure in a condition known as hydrostatic equilibrium. We know that, for some clouds, gravity *must* be the confining force, for it is gravity that causes the contraction of clouds into new stars.

A self-gravitating cloud, however, must have enough mass and a certain density to overcome the outward pressure. For more diffuse clouds, gravity is too weak to provide cohesion. These clouds must be confined by a surrounding gas, an intercloud medium whose own counteracting pressure maintains their integrity. Such a medium must be even more tenuous than the clouds or it would be as obvious as the clouds themselves. Since pressure increases with temperature, the confining medium must be warm, even hot. The presence of such a gas is exposed by a variety of observations and measures.

We see three components. A *warm neutral medium* is revealed by 21-cm emission. An emission line coming from an interstellar cloud really represents a blend of the photons produced by the individual atoms. Because of the cloud's temperature, each atom is moving at a slightly different velocity. Some atoms are approaching while emitting, others receding. As a result, each photon is slightly Doppler-shifted relative to the average for the cloud, and the resulting line is thermally broadened. It may be further broadened if there are turbulent motions within the cloud. The cool H I clouds have a low degree of broadening, consistent with low temperature. However, radio astronomers also see superimposed on the spectra of the individual clouds a much broader spectrum line that implies temperatures of some 8000 K. This broad line is not seen in absorption against bright sources, and it is therefore not coming from small discrete clouds, but from much larger volumes of very low density, averaging some 0.5 atoms per cubic centimeter.

In regions of space not shielded from that portion of the radiation of hot O stars that escapes the diffuse nebulae, we find a *warm ionized medium,* visible through its faint extensive Hα radiation. It is rather like a thin but huge diffuse nebula that pervades much of

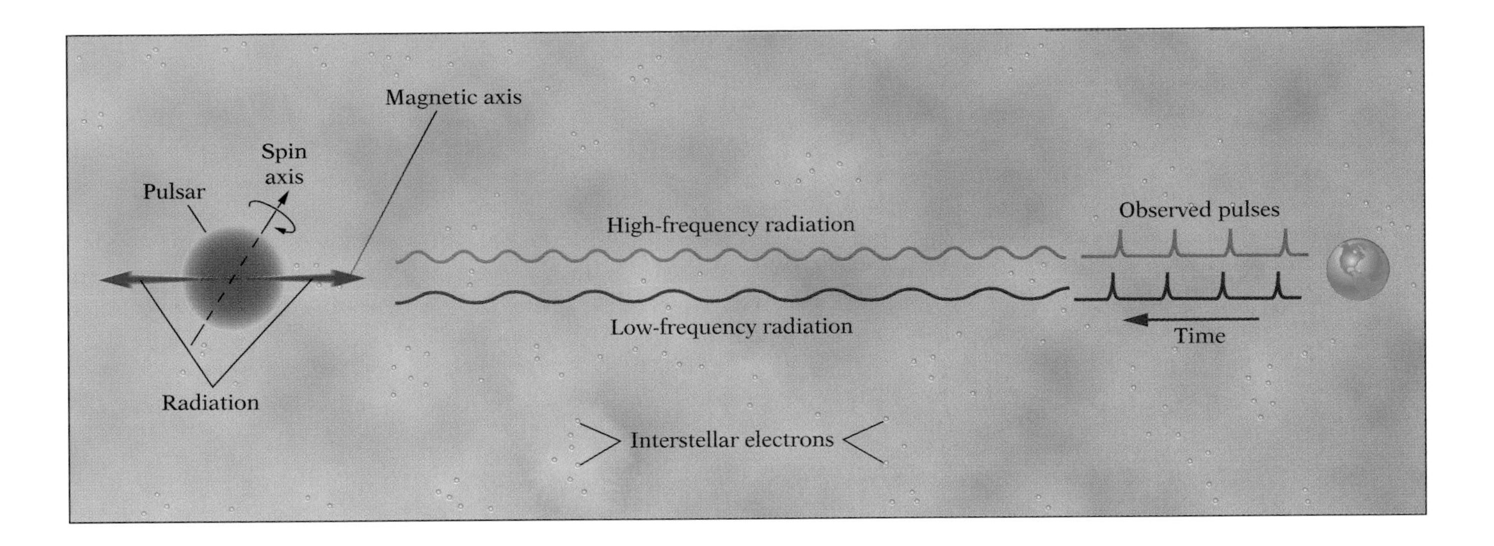

Low-frequency radio radiation from a typical pulsar, a rotating neutron star, arrives behind high-frequency radiation because of refraction by electrons in the warm ionized interstellar medium.

the Galaxy. Faint forbidden radiation implies, again, temperatures of about 8000 K. In the wonderful synergy of astronomy, it is best explored by means of the dense dregs of stellar evolution called neutron stars. O stars, those that produce the diffuse nebulae from the dark clouds, ultimately develop iron cores through continuing internal thermonuclear fusion. When, at the end of the fusion chain, the cores collapse, these stars explode in grand, brilliant "supernovae." The expanding debris of the exploded stars becomes "supernova remnants" (to be explored in Chapter 9); their shock waves are believed to produce the interstellar diamonds from ordinary carbon dust. The ex-iron cores, however, are so dense they cannot fly apart: the protons and electrons merge, the stellar remainders becoming neutron stars no more than 30 km or so across, with extraordinary average densities of 10^{14} cm^{-3}, and magnetic fields condensed to a trillion or so times that of the Earth.

Newly minted neutron stars spin quickly, up to 30 times per second. The magnetic field causes radiation to be funneled outward along a tilted magnetic axis. As the axis sways in space, the twin beams wobble like a lawn sprinkler. If the Earth is in the way, we see a pulse of radiation and deem the otherwise invisible object a pulsar. Over 600 are known, most pulsing more slowly and only in the radio spectrum.

All the radio frequencies emitted by the source are produced at about the same time. However, on Earth we receive high-frequency pulses *first*. Light, radio, and the rest of the electromagnetic spectrum travel at c, the "speed of light," only in a vacuum. When electromagnetic waves travel through a substance, they are slowed down. Slowing causes the refraction, or bending, of light, an effect we see when light passes through water or glass, distorting the shape of fish in an aquarium. Radio waves are refracted by their passage through an ionized gas. Along with refraction goes dispersion, a change of speed with frequency, in which lower-frequency waves are slowed the more. The variation in pulse arrival times with frequency therefore implies free electrons in interstellar space and thus a warm ionized medium with an electron density of about 0.15 per cubic centimeter.

New technologies invariably result in new and unexpected discoveries. The *Copernicus* satellite, which allowed detailed examinations of ultraviolet stellar spectra, revealed interstellar absorption lines from highly ionized atoms, particularly the O VI resonance line (from O^{+5}) at 1035 Å. To obtain an ionization level this high requires enormous temperatures, in the hundreds of thousands of kelvins. Even earlier orbiter missions had detected diffuse X-ray emission from space, that is, X rays that did not have a discrete source and that could only be produced by a diffused gas in the neighborhood of a *million* kelvins. Interstellar space therefore also contains a *hot ionized component,* called the coronal gas or coronal component in analogy to the hot corona of the Sun. The temperature of the coronal gas is much too high to be produced by absorption of radiation from hot stars (the gas could not have a temperature hotter than the hottest stars). The only source known to be available for such heat is shock waves from exploding stars, from supernovae. Thus the supernovae, whose leavings, the neutron stars that allow the examination of the warm ionized gas, in themselves produce yet another component of the interstellar medium. As we will see, supernovae have other profound effects, perhaps even contributing to the formation of the Earth itself.

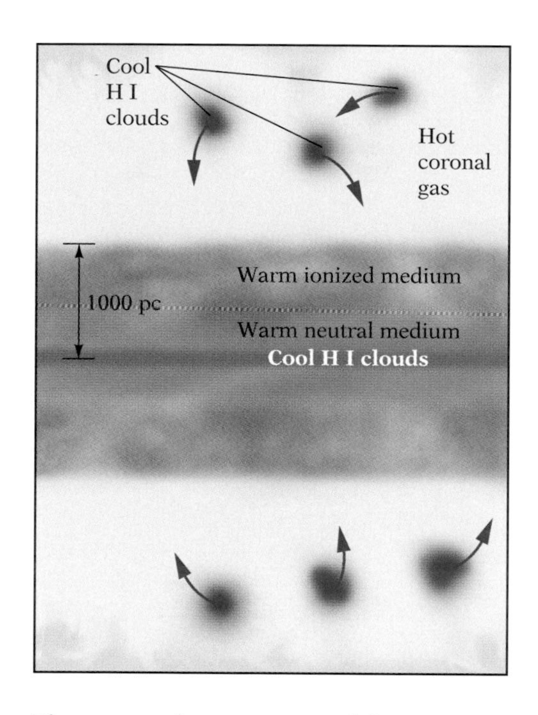

The warmer the component of the interstellar medium, the more it expands away from the midplane of the Galaxy. The cool H I clouds are generally confined to a very thin layer, but a few are seen at high elevations above (and below) the disk. The warm neutral medium and the warm ionized medium lie in much thicker layers, and the coronal gas billows to great heights. The hot component may be fountaining into the halo in part as a result of supernova explosions, the cooling gas then flowing back into the disk.

The interrelations among all these components are complex and still not very well understood. The pressures of the warmer components seem to help hold together the diffuse neutral clouds, those not confined by gravity; true "pressure equilibrium," however, is probably not close to being achieved, allowing continuous changes in the structure of the interstellar medium. We have yet to determine the volume of space and the fraction of interstellar mass occupied by each constituent. Part of the problem is that the interstellar components are not cospatial—that is, they do not occupy the same places within the Galaxy's structure. The cool neutral clouds, like the dense globules, occupy a thin disk only a couple hundred parsecs thick, extending to about 100 pc above and below the plane. There are no sharp boundaries, however. The so-called edge is described by a scale height at which the spatial density of the clouds (the number per unit volume) drops to 1/e of its central value (where e, the base of the natural logarithms, equals 2.714 . . .). Some neutral hydrogen clouds violate even this simple description, as we find renegades at quite high galactic latitudes a few thousand parsecs off the midplane, some of them with anomalously high radial velocities. Heated air expands: gently warm a balloon and watch it swell. The warmer components of the interstellar medium therefore expand well beyond the midplane. The scale heights of the warm neutral and warm ionized media are 500 pc and 1000 pc respectively, and that of the coronal gas is 3000 pc, taking the hot interstellar medium well into the galactic halo.

The cool H I clouds hold, *very* roughly, about a quarter of the interstellar medium's mass, but because they are relatively dense they occupy only about 2 percent of the volume. The two warm components hold more mass, perhaps half the total, two-thirds of it in the neutral phase; and together they occupy half the volume as well. The coronal gas fills perhaps as much as half the remaining volume, but because of its extremely low density—only 10^{-3} atoms cm^{-3}, consistent with its high temperature—contains only a few percent of the mass.

The whole system is much more complex than this simple description implies. The interstellar medium is a dynamic, constantly changing structure. The density waves that produce the spiral arms act like giant pistons that compress the interstellar gases, helping to create the clouds, which constantly collide, break apart, and reform. The cool "clouds" are in fact probably not really clouds at all, but sheets and filaments, some—many of them—at the edges of expanding hot bubbles made by the blast waves of supernovae. The interstellar medium may have a frothy kind of structure, like a sponge, in which hot expanding supernova bubbles weave their way through the cooler gas. A holistic view even likens the interstellar medium to a fountain in which hot gas generated in supernova explosions is lofted far into the galactic halo, where it cools into denser neutral high-velocity clouds that rain back into the disk, to be thrown up again in a perpetual cycle.

This increasingly complex picture does not factor in the energy of the galactic magnetic field, which may be critical in supporting the hot gas at high galactic elevations (relieving the supernovae of that role), nor of cosmic rays, accelerated high-speed atomic nuclei, that can deposit considerable additional energy. All in all, though astronomers have made startling advances in our knowledge of the interstellar medium over the past few decades, uncertainty still reigns.

The Local Medium

We tend perhaps to look at the interstellar medium as "out there," having little to do with us. But the Sun orbits the Galaxy within the medium; it is part of our milieu. The diffuse X rays are readily absorbed by interstellar gas, so they must be produced close to us. The interstellar O VI absorption lines are also found in stars that lie quite nearby. Our system therefore seems to lie principally within a bubble of hot coronal gas, the local bubble, which was probably produced relatively recently by a supernova. Radio observations at 21 cm indicate little neutral gas, though optical observations of interstellar lines do reveal a few neutral clouds within a couple hundred parsecs of us. Ultraviolet interstellar absorptions suggest that we are actually inside and near the edge of a tenuous neutral cloud that extends perhaps 10 pc across within the local bubble. This and the other clouds seem to be blowing at a speed of a few kilometers per second out of the direction of Scorpio and Centaurus, regions that contain a large number of massive young O and B stars, the clouds perhaps driven by the explosions of supernovae.

Remarkably, we—humanity—are about to enter the local interstellar medium. The Sun continuously loses mass at a very low rate, about 10^{-13} solar mass per year. This solar wind, blowing past us at a speed of a few hundred kilometers per second, is ultimately responsible for the northern lights and the streaming gaseous comet tails that always point away from the Sun. The solar wind keeps the local interstellar medium at bay, blowing a small bubble (vastly smaller than the local bubble) within it. Four spacecraft—the twin *Pioneer*s that observed Jupiter and Saturn in the early 1970s and the famed *Voyager*s—have left the planetary system. The *Voyager*s should have enough on-board power to last until, sometime early in the twenty-first century, they cross the boundary of the bubble blown by the solar wind, a shock wave known as the heliopause.

We already have indications of the heliopause from a nine-month-long burst of radio radiation received by the *Voyager*s in 1992 and 1993. The onset took place about a year after an intense period of solar flaring and coronal outbursts caused by the release of solar magnetic energy associated with the Sun's 11-year magnetic activity cycle. We surmise that the resulting shock wave raced along the solar wind, eventually striking and exciting the heliopause and generating the observed radiation. From the known speed of such disturbances, we estimate the heliopause to be somewhere between 100 AU and 150 AU from the Sun. We are about to cross the great divide and step into true interstellar space.

Look again at the larger picture, the setting of the local bubble. The mass fractions of the interstellar medium examined so far —about a quarter for the H I clouds and half for the intercloud medium—do not add to 1. Something big is missing. A portent of this critical gap was revealed in the late 1930s by the interstellar absorption lines of the simple molecules CH and CN. They were the vanguard of data from perhaps the most important part of interstellar space, certainly the most directly important to us, the cold molecular clouds. One of them gave us birth.

5

MOLECULES!

≺ *Bucky balls, C_{60} molecules, epitomize the complexity of interstellar chemistry.*

A deluge begins with a single drop, then another, then more and more until there is a drenching cascade. After their discovery, the interstellar molecular absorption lines of CH, CH^+, and CN lay quietly—and unaccompanied—for some 25 years. Most astronomers were fairly well convinced that interstellar molecules were not very important—difficult to make in the low densities and temperatures of space, and easily destroyed by high-energy stellar radiation. A few prescient theorists listed the lines of some molecules that might be observable in the radio spectrum, and in 1963 a radio telescope pointed toward Cassiopeia A detected four absorption lines, identified with hydroxyl (OH), against its bright background. Five years after that, ammonia (NH_3) emissions were seen in the direction of the center of the Galaxy. Water was found a year later, and the breathtaking discovery of formaldehyde (H_2CO), an *organic* (carbon-containing) molecule, soon followed. New molecules tumbled in, including carbon monoxide (CO), molecular hydrogen (H_2, actually first observed in the ultraviolet), hydrogen cyanide (HCN), and—flowing from the taps of the celestial tavern—ethyl alcohol (CH_3CH_2OH).

Until the 1970s, astronomy by tradition and necessity had concentrated on physics, not chemistry—stars, after all, are mostly too hot to contain much in the way of molecules. The interstellar medium taught us differently. Though 90 interstellar molecules and molecular ions are now known, astronomers have only begun to explore a chemistry so rich it may even contain the seeds of life.

COMBINING THE ATOMS

All the Universe is made from a few simple particles of normal matter—the protons, neutrons, and electrons that compose atoms. Arranged in different ways, they produce the chemical elements. A few elements are commonly encountered in their pure state: carbon in a diamond ring, iron in a fence, lead in a fishing sinker, zinc on a galvanized garbage can. But the vast majority of substances are combinations of atoms that make a near-infinite variety of molecules that do not carry the properties of the atoms that make them but are entirely new species with totally different characteristics. Chlorine (atomic number 17) is highly toxic, but combine it with sodium (number 11) and you can salt your food.

To see how molecules are built, we need go back to the atom. Normal, neutral, hydrogen has one negative electron to balance its single nuclear positive proton. Helium, atomic number 2, normally (unless it is ionized) has two electrons. In the ground state—that is, unless an electron is excited to a higher orbit or energy level—the two electrons have the same orbital radii. Pass on to lithium and you add a third electron—but this one finds itself frozen out of the brotherhood of the first orbit.

Atomic particles—here, specifically the electrons associated with atoms—are described in quantum mechanics by four quantum properties, each of which is associated with a value, a quantum number. Two of these were encountered in the description of the simple Bohr atom: the principal quantum number that describes orbital sizes and the angular momentum number related to orbital eccentricities. The third, which relates to spin, is needed to describe the 21-cm radio line and can take on only two values, +1/2 and –1/2. The fourth is a "magnetic" number related to the angular momentum number. The Pauli exclusion principle (named after Wolfgang Pauli, a theoretician so "pure" it was said his mere presence in a town would cause experiments to go awry) states that no two electrons (or similar particles like protons and neutrons) can have identical quantum numbers. Because helium's two electrons have opposite spins, they are not identical, and therefore they can coexist, occupying a "shell," which is now closed to other electrons. To

The periodic table organizes the known elements. Colored boxes denote those with only unstable—radioactive—isotopes; those in red are too unstable to exist naturally on Earth. Each row shows another shell of electrons, and each column shows elements with similar chemical properties.

14 — Atomic number
Si — Symbol
Silicon — Name
28 — Atomic weight of most common isotope

1 H Hydrogen 1																	2 He Helium 4
3 Li Lithium 7	4 Be Beryllium 9											5 B Boron 11	6 C Carbon 12	7 N Nitrogen 14	8 O Oxygen 16	9 F Fluroine 19	10 Ne Neon 20
11 Na Sodium 23	12 Mg Magnesium 24											13 Al Aluminum 27	14 Si Silicon 28	15 P Phosphorus 31	16 S Sulfur 32	17 Cl Chlorine 35	18 Ar Argon 40
19 K Potassium 39	20 Ca Calcium 40	21 Sc Scandium 45	22 Ti Titanium 48	23 V Vanadium 51	24 Cr Chromium 52	25 Mn Manganese 55	26 Fe Iron 56	27 Co Cobalt 59	28 Ni Nickel 58	29 Cu Copper 63	30 Zn Zinc 64	31 Ga Gallium 69	32 Ge Germanium 74	33 As Arsenic 75	34 Se Selenium 80	35 Br Bromine 79	36 Kr Krypton 84
37 Rb Rubidium 85	38 Sr Strontium 88	39 Y Yttrium 89	40 Zr Zirconium 90	41 Nb Niobium 93	42 Mo Molybdenum 98	43 Tc Technetium 99	44 Ru Ruthenium 102	45 Rh Rhodium 103	76 Pd Palladium 106	47 Ag Silver 107	48 Cd Cadmium 114	49 In Indium 115	50 Sn Tin 120	51 Sb Antimony 121	52 Te Tellurium 130	53 I Iodine 127	54 Xe Xenon 132
55 Cs Cesium 133	56 Ba Barium 138	57 La Lanthanum 139	72 Hf Hafnium 180	73 Ta Tantalum 181	74 W Tungsten 184	75 Re Rhenium 187	76 Os Osmium 192	77 Ir Iridium 193	78 Pt Platinum 195	79 Au Gold 197	80 Hg Mercury 202	81 Tl Thallium 205	82 Pb Lead 208	83 Bi Bismuth 209	84 Po Polonium 210	85 At Astatine 210	86 Rn Radon 222
87 Fr Francium 223	88 Ra Radium 226	89 Ac Actinium 227	104 261	105 262	106 263	107 262	108	109	110								

Metals | Nonmetals

Lanthanide Series

58 Ce Cerium 140	59 Pr Praseo-dymium 141	60 Nd Neodymium 142	61 Pm Promethium 145	62 Sm Samarium 152	63 Eu Europium 153	64 Gd Gadolinium 158	65 Tb Terbium 159	66 Dy Dysprosium 164	67 Ho Holmium 165	68 Er Erbium 166	69 Tm Thulium 169	70 Yb Ytterbium 174	71 Lu Lutetium 175
90 Th Thorium 232	91 Pa Proatactinium 231	92 U Uranium 238	93 Np Neptunium 237	94 Pu Plutonium 242	95 Am Americium 243	96 Cm Curium 247	97 Bk Berkrlium 249	98 Cf Californium 251	99 Es Einsteinium 254	100 Md Mendelevium 256	101 Fm Fermium 253	102 No Nobelium 254	103 Lr Lawrencium 257

Actinide Series

prevent identical quantum numbers, lithium's third electron takes on a new principal number: that is, it has a larger orbital radius, creating a new shell.

The design of the periodic table, the organization chart of the atoms, depends on shell closures. Helium, whose two electrons fill the first shell (with principal quantum number 1), ends the first row of the table. Larger shells have greater capacities (each orbit n has n eccentricity states); they can hold greater numbers of electrons without there being identical quantum numbers. As we move through the elements, adding protons through beryllium (atomic number 4), boron (5), carbon (6), nitrogen (7), oxygen (8), and fluorine (9), we fill the second shell, which closes at neon (10), at the end of the periodic table's second row. The next element, sodium (atomic number 11), adds its eleventh electron in a *third* shell, which can contain another seven electrons. This shell fills at argon, completing the third row of the periodic table. Though higher atomic numbers lead to complications wherein new shells open before older ones are closed (resulting in the outlander "lanthanide" and "actinide" rows of the periodic table), there is a strict order to it all.

Electrons in unfilled outer shells, the so-called valence electrons, can be shared among atoms, thus locking them together—in other words, atoms can combine to form molecules. In this simple statement lies all the richness of the natural world. In the simplest case, two identical atoms bond into their diatomic form: two hydrogens make molecular hydrogen, H_2; two oxygens make molecular oxygen, O_2. In spite of the similarity of the names, the molecular forms are very different from the atomic. The air we breathe contains molecular nitrogen, N_2, into which is mixed life-giving O_2; atomic oxygen, however, is poisonous. Diatomic molecules can also be made from dissimilar atoms—common examples are CO (carbon monoxide), HCl (hydrogen chloride), and NaCl (table salt).

Molecules bind best when the valence electrons fit. Lithium has one outer electron. Fluorine needs one electron to make a filled shell; that is, it has a single "hole" in its outer electron shell. The two atoms thus combine violently to make LiF, lithium fluoride (do not try this experiment at home). The columns of the periodic table contain atoms with the same number of valence electrons: those at the left end have one valence electron (the start of a new shell); those at the right have none—these shells have all the electrons they can take, with no holes to be filled. The atoms in the same column therefore have similar chemical properties, forming into molecules in similar (though certainly not identical) ways. Sodium and chlorine combine in the same way as do lithium and fluorine, just not quite as explosively (don't try this either). Atoms bind the worst when their outer shells are filled. Helium, at the end of a row of the periodic table, needs no partner and indeed will not take any. Nor will neon, argon, xenon, or radon, the set called the noble gases because they always stand alone (actually, molecules *have* been made of them, but not easily).

Unbalanced electron pairings make unstable molecules known as radicals. Oxygen needs two electrons to make a filled shell, but a single hydrogen atom can supply only one, so hydroxyl (OH), a prime example of a radical, will search for another H atom to make a triatomic molecule, H_2O—water—which is highly stable. Although hard to make on Earth, radicals can exist in the interstellar medium in abundance because the extremely low densities and low temperatures make it unlikely that they will encounter another atom or be collisionally disrupted.

Combinations of three atoms like water, carbon dioxide (CO_2), and nitrous oxide (N_2O) vastly increase the number of possibilities. With four (ammonia, NH_3), five (methane, CH_4), six, seven, and more atoms, the possibilities of combinations concomitantly increase yet further. Carbon holds a special place: with four valence electrons it can combine in a vast number of ways, tying up a variety of other atoms into an uncountable number of enormously complicated molecules soap has dozens of atoms, and deoxyribonucleic acid (DNA) has hundreds of thousands. (Silicon plays a similar, though much lesser, role in the next row of the periodic table.) Carbon molecules form the basis of life and thus are termed organic whether they actually are involved with life or not (in spite of your mother's instructions to wash with soap before dinner, soap is not necessary to life). Molecules without carbon are thus *inorganic* even if, like water, they *are* needed for life (showing that it is not just astronomy that has an idiosyncratic vocabulary). Organic chemical formulas commonly reflect the order in which the atoms are combined along the length of the molecule. Methyl formate, for example, is $HCOOCH_3$, not $HC_2O_2H_3$. Like atoms, molecules can be ionized: CH (a radical, as is CN) when missing an electron becomes the CH^+ also found in interstellar space. Because the ions have different chemical properties, they are, astronomically anyway, considered different species.

The ties that bind atoms into molecules are generally weaker than those that bind electrons to nuclei. Energies (of collision or radiation) that would merely excite electrons to higher energy levels can easily dissociate molecules, tearing them apart. As a result,

The great Ophiuchus dust lanes, stretching above and to the left of the bright stars in Scorpius (near the center of the picture), are cold chemical factories in which vast numbers of molecules have been formed. They are also breeding grounds of stars.

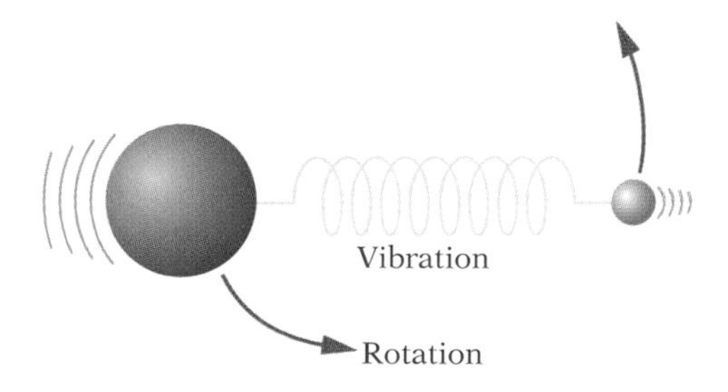

In addition to their electronic energies (not shown here), molecules carry quantized energies of rotation around a common center of mass and a quantum-mechanical version of springlike vibration along an axis.

molecules cannot exist at high temperatures. In the solar atmosphere, for example, only molecules that, like CH, are tied tightly enough to withstand the constant collisions of a 6000-K gas can survive. At 9000 K, the surface temperature of the bright star Vega, no molecules are left at all. In the lower stellar temperature reaches, however, molecular astronomy becomes quite important; cool stellar spectra are dominated by fragile titanium oxide (TiO). Carbon stars (those that, for evolutionary reasons, have more carbon than oxygen, the reverse of the normal situation) are loaded with C_2, CN, and many other compounds. In the coolest stars we also see oxides of other metals and even water vapor.

In spite of the early observations of CH and CN, no one thought space chemistry would be very significant. The low densities and temperatures of the interstellar medium were expected to inhibit molecule formation, and high-energy photons from hot stars should quickly destroy those that could be created. Now, however, consider the interstellar refrigerator. Dusty clouds block high-energy stellar radiation, and, without an external source, they are cold. Thus impact energies are low and also, because of low densities, infrequent. Furthermore, with no high-energy photons to disrupt the molecules directly, molecules once made can survive in abundance.

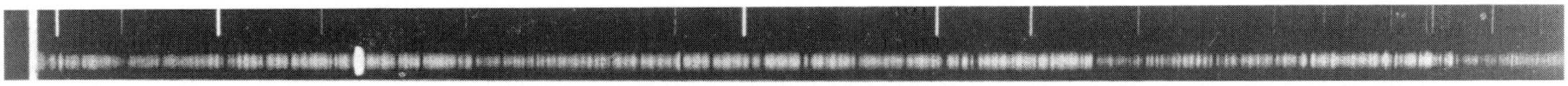

The spectrum of the star Omicron Ceti —Mira—appears almost hopelessly complex, so cut up by molecular bands, mostly of TiO, that the continuum disappears, the lines severely overlapping.

The real inhibition to molecular astronomy, though, was observational. Low temperatures align with low-energy radiation, not with that in the optical domain. Astronomers needed a new way of looking, and in 1963 they turned the radio on.

SPREADING THE SPECTRUM

Discovery, however, requires spectra, and you cannot beat molecules for spectral interest and variety. Compared with them, the spectra of even the heaviest atoms are simple. All atomic spectra are caused by electronic transitions between various energy levels. Molecules, however, have much more freedom and can carry energy by rotation and vibration. A simple diatomic molecule classically behaves something like a dumbbell spinning around an axis perpendicular to the line joining the atoms, each "weight" vibrating as if it rode the end of a connecting spring. Like electronic transitions, rotation and vibration are quantized—that is, rotation and vibration can take on only certain energies. When these quantized energies are added to the electronic energies, the resulting spectral complexity is awesome.

Place a simple diatomic molecule in its ground state, where the electrons have their lowest energies, and the molecule sits quietly, neither vibrating nor rotating. Now set it rotating. The lowest energies above the ground state are reached as the molecule spins at ever increasing, though quantized, speeds. The energy levels are initially closely spaced, but they spread out as centrifugal force stretches the bond. Transitions, permitted only between adjacent states, have very low energies and are commonly found in the very short wave radio—the microwave—part of the spectrum. The wavelengths of their photons are typically measured in millimeters (hence the need for a radio telescope) or in the long-wave end of the infrared, the "far" infrared. Permitted rotational spectra, however, are present only when the two atoms of a diatomic molecule are different, as in CN or CO. If they are the same, as in H_2 or C_2, the transitions are forbidden (in the same sense as the "nebulium" lines) and the lines are relatively weak.

Now still the rotation and allow the molecule only to vibrate. The vibrational energy levels ascend more like those of hydrogen,

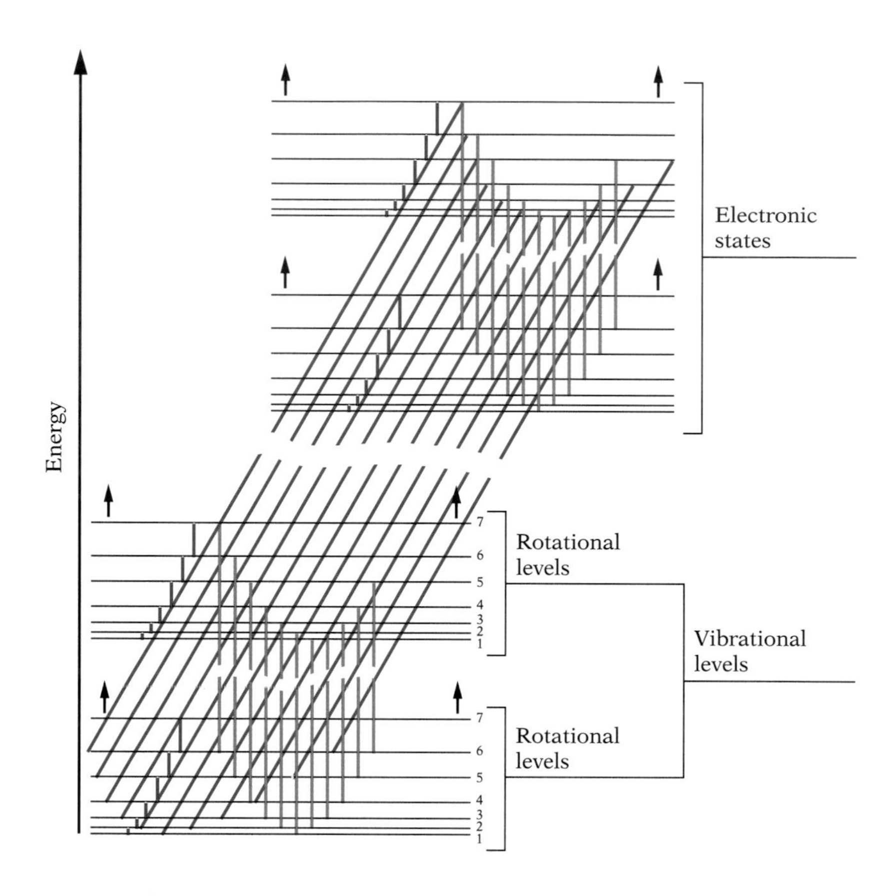

The electronic states of molecules are broken into vibrational levels (only two of which are shown here), which in turn are divided into rotational levels (only seven of which are shown). Pure rotational (low-energy), rotational–vibrational (medium-energy), and electronic–rotational–vibrational (high-energy) transitions are possible. Each electronic state has many more vibrational levels, and each vibrational level many more rotational levels, than shown. Transitions usually take place between adjacent rotational levels, but other transitions are possible.

getting closer and closer together as energy increases. At low energies, the vibration is close to linear—that is, the atoms are displaced to equidistant sides of the rest position. But as energy increases, the outward part of the vibration moves farther from the rest position than the inward: the inward vibration can only go so *far* inward, but the outward has no bound. Eventually, the atoms are so far apart that the bond breaks and the molecule dissociates in a way analogous to ionization. Transitions between adjacent vibrational states are by far the most likely at low energies, but at higher energies larger jumps are quite possible. Since vibration energies are greater than those of rotation and the energy levels farther apart, the transitions between the vibrational states emit or absorb radiation at higher frequencies (and lower wavelengths) and are present throughout the infrared into the millimeter radio.

Next, combine the two and set the molecule rotating and vibrating at the same time. Each vibrational level is now broken into rotational sublevels. Transitions can go from one rotation–vibration state to another as long as the change in rotation is one quantum unit. There is now an immense number of possibilities, the sum of all transitions arrayed in a pair of bands, each of which has a huge number of lines that can converge on a "band head," or limit.

Finally, add the electronic states. Each has a set of vibrational states, and each of these is further broken into rotation states. Electronic transitions now produce not just a single line but a whole system of bands. Electronic energies are higher than either rotation or vibration energies alone, and excited electronic bands appear throughout the optical into the ultraviolet. In the spectrum of a cool star, several systems of bands may overlap, each produced by a different transition or molecule.

And all this complexity comes from just *diatomic* molecules. Polyatomic molecules are more complex yet because they have modes of rotation and vibration along each of their different axes; and they may be asymmetric as well, adding further to the complexity. Fortunately, for all the intricacy of molecular spectra, astronomers usually deal with only a few well-defined and -understood bands or individual vibrational or rotational lines. Their study throws open the door to interstellar space and star formation.

AN ASTONISHING VARIETY SHOW

Interstellar atomic lines offer little in the way of surprises, the greatest perhaps being the depletions of the gaseous elements and supposed accretions onto dust grains. Not so the molecular lines. The first radio molecular lines, those of the OH radical, were discovered in the 2-cm band, the observed frequencies of 1612 to 1720 MHz identified with the hyperfine structure of the lowest rotation–vibration level. Two years later, in 1965, the OH lines were detected in emission from regions rich in small, diffuse nebulae. However, their intensities were not in the ratios expected had they been emitted by a "thermal" gas, one at equilibrium, whose molecules and atoms are excited by collisions with particles moving with a Maxwellian velocity distribution. The intensities of a pair of OH lines are extraordinary, requiring a temperature of 10^{18} K, absolutely impossible within interstellar space, absolutely impossible *anywhere*. Such lines had to have been amplified: they could have been produced only by a natural interstellar maser. Cold interstellar space was about to become one of the hottest topics in astronomy.

The maser—*m*icrowave *a*mplification by the *s*timulated *e*mission of *r*adiation—had already been discovered in the laboratory and

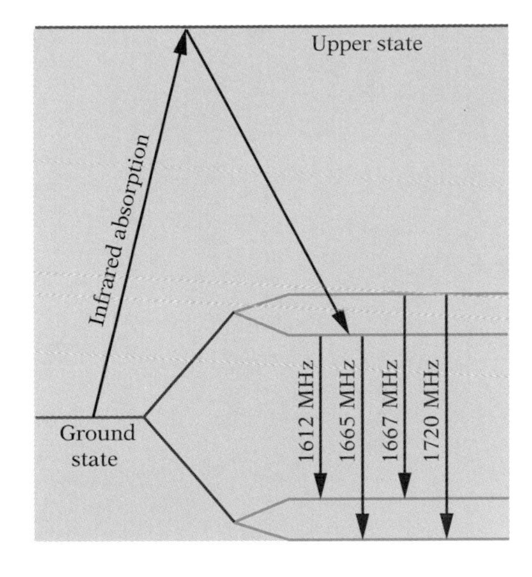

The ground state of OH is doubled (red), then each level is split into a hyperfine structure (green); the result is four microwave lines. The third hyperfine state receives electrons from a higher state that is excited by infrared radiation. As a result, it has far too many electrons relative to equilibrium, the condition required for a natural maser, and the lines at 1612 and 1655 Mhz are pumped to extraordinary luminosities.

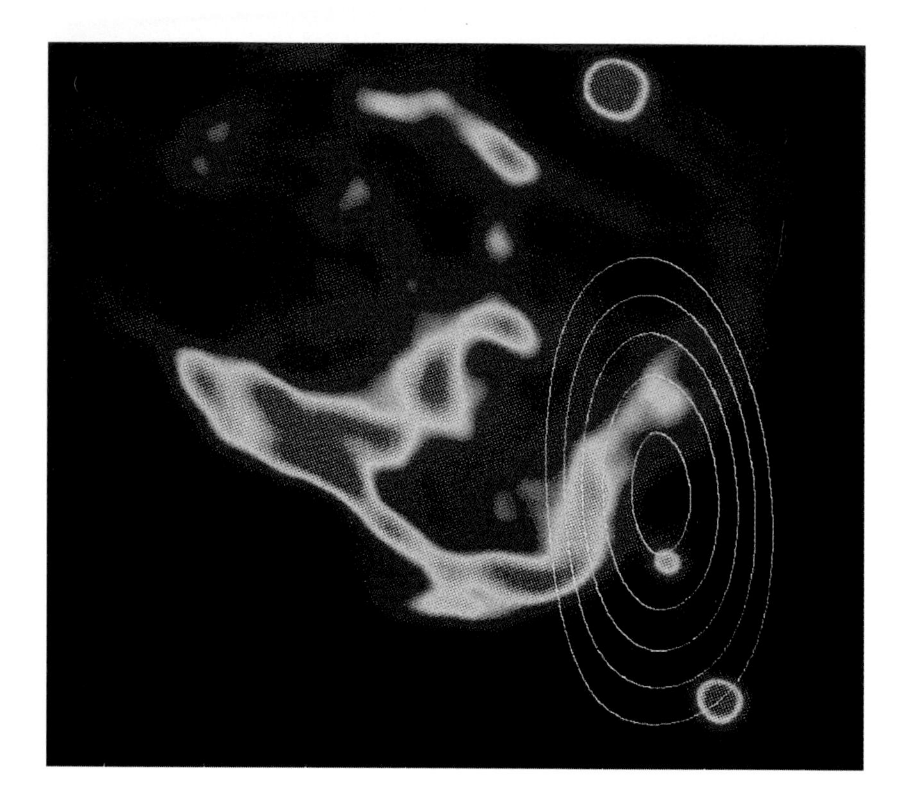

Sagittarius B2 North is a part of a molecular cloud complex near the center of the Galaxy that contains a small source of radiation so extraordinarily rich in complex molecules that it is called the "large molecule Heimat (home) source." The colored region shows continuous radio (free-free) radiation from associated diffuse nebulae; the contours show the location of acetone, $(CH_3)_2CO$, in the molecular gas.

was in fact being used by radio astronomers to amplify faint celestial signals. Its optical counterpart, the laser—*l*ight *a*mplification by the *s*timulated *e*mission of *r*adiation—is now commonplace, used in door openers, supermarket scanners, weapons, and surgical instruments. But the discovery of masering action *in nature* was amazing.

The process that allows the creation of the maser or laser had been known for decades. Radiation and atoms (and of course molecules) interact in three ways: absorption, *spontaneous* emission, and *stimulated* emission. In spontaneous emission an electron jumps downward from an upper orbit or energy level to a lower one all by itself. In stimulated emission the electron has help. As in absorption, the incoming photon must have an energy equal to the energy difference between the two levels. Instead of disappearing while raising an electron upward, the photon makes an electron in the upper state drop *downward* without damage to itself. As a result, two photons fly off in the same direction in phase with each other, their waves oscillating together. There is nothing strange about stimulated emission. In any thermal gas interacting with radiation all three processes occur at once, and all three, along with collisions, are responsible for the level populations (the numbers of electrons in given levels), which normally decrease sharply as energy increases.

A maser or laser is produced when the level populations are distorted, making an upper level more populous than it would be in an ordinary thermal gas—that is, it must be pumped up, even inverted, so that there are more electrons in an *upper* level than in a *lower.* Since the level populations are inverted, the stimulated emissions dominate over the inevitable absorptions. One photon of the right wavelength, either emitted spontaneously or coming in from outside, then causes another electron to drop; the resulting two photons generate two more, each of these two more, until there is a chain reaction of dropping electrons and emitted photons, a cascade, a storm, of them. The final result is a massive amplification of the beam and a powerful, narrow, extremely intense ray of radiation. The OH masers are small, dense clouds in which the molecules are pumped up by the infrared light radiated by dust surrounding embedded newly forming stars. This process is another in the growing list of phenomena that relate the interstellar medium, and dense clouds, to star formation.

In spite of its remarkable masering properties, OH is nonetheless a simple diatomic molecule like CN, CH, and CH^+. Interstellar chemistry at best still appeared relatively boring. In the face of skepticism that polyatomic molecules could exist, C. H. Townes, of

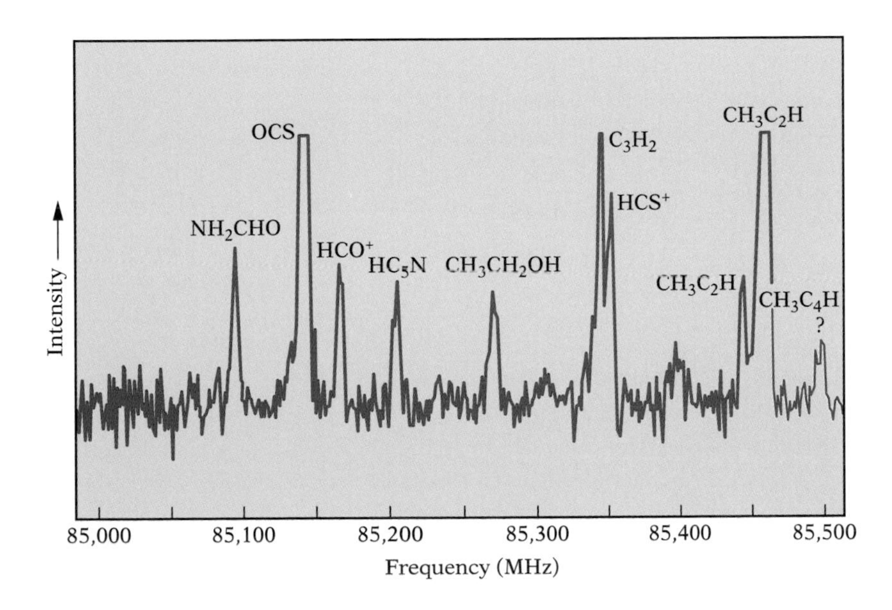

The radio spectrum of Sagittarius B2 displays the cloud's great organic riches.

A Sample of Interstellar Molecules

Stable Inorganic Molecules

Diatomic		Triatomic		Four-Atom	
H_2	hydrogen	H_2O	water	NH_3	ammonia
CO	carbon monoxide	H_2S	hydrogen sulfide		
CS	carbon monosulfide	SO_2	sulfur dioxide	Ammonia	
NO	nitric oxide	OCS	carbonyl sulfide		
NS	nitrogen sulfide	HNO	nitroxyl		
SiO	silicon monoxide	C_2O	dicarbon monoxide		
SiS	silicon sulfide	C_2S	——		

Stable Organic Molecules

Alcohols		Aldehydes and Ketones		Acids		Hydrocarbons	
CH_3OH	methanol	H_2CO	formaldehyde	HCN	hydrocyanic	C_2H_2	acetylene
CH_3CH_2OH	ethanol	CH_3CHO	acetaldehyde	HCOOH	formic	CH_4	methane
		CH_2CO	ketene	HNCO	isocyanic	Acetylene	
				CH_3COOH	acetic		

Amides		Esters and Ethers		Organo Sulfur		Biomolecules	
NH_2CHO	formamide	HCOOCH	methyl formate	H_2CS	thio-formaldehyde	NH_2CH_2COOH	glycine[a]
NH_2CN	cyanimide	CH_3CH_3O	dimethyl ether	HNCS	isothiocyanic acid	Glycine	
NH_2CH_3	methylamine			CS_3SH	methyl mercaptan		

[a]Tentative identification.

Source: Adapted from B. Turner, "Interstellar Medium, Molecules," in *The Astronomy and Astrophysics Encyclopedia,* Stephen P. Maran, ed. (New York: Van Nostrand Reinhold, 1992; modified from E. F. van Disheok et al., *Protostars and Planets III,* Eugene H. Levy and John I. Lunine, eds. [Tucson: University of Arizona Press, 1993]). Only confirmed true *interstellar* molecules are included here; a variety of molecules found in envelopes around stars, *circumstellar* molecules, are not included. This table does not include possible large carbon molecules such as PAHs and buckminsterfullerenes.

Paraffin Derivatives		Acetylene Derivatives		Other	
CH_3CN	methyl cyanide	HC_3N	cyanoacetylene	CH_2NH	methylenimine
CH_3CH_2CN	ethyl cyanide	CH_3C_2H	methylacetylene	CH_2CHCN	vinyl cyanide

Unstable Molecules

Radicals				Ions	
CH	methylidine	C_3N	cyanoethynyl	CH^+	methylidine ion
CN	cyanogen	C_3O	tricarbon monoxide	HCO^+	formyl ion
OH	hydroxyl	C_4H	butadinyl	HOC^+	isoformyl ion
SO	sulfur monoxide	C_5H	pentynylidyne	$HOCO^+$	protonated carbon dioxide
HCO	formyl	C_6H	hexatrinyl	HCS^+	thioformyl ion
C_2H	ethynyl	C_2S	dicarbon sulfide	H_3O^+	hydronium ion
C_3H	propynylidyne	CH_2CN	cyanomethyl	$HCNH^+$	protonated hydrogen cyanide

Rings		Carbon Chains		Isomers	
C_3H_2	cyclopropenylidene	C_3S	tricarbon sulfide	HNC	hydrogen isocyanide
C_3H	propynylidyne	HC_5N	cyanodiacetylene	CH_3NC	methyl isocyanide
		HC_7N	cyanohexatriyne	HCCNC	isocyanoacetylene
		HC_9N	cyanooctatetrayne		
		CH_3C_3N	methyl cyanoacetylene		
		CH_3C_4H	methyl diacetylene		
		CH_3C_5N	——		

Cyclopropenylidene

the University of California, predicted the frequencies of ammonia (NH_3) lines, and in 1968 he and his group found ammonia in emission toward the center of the Galaxy. (A great number of ammonia lines are now seen in both emission and absorption.) Then came the discoveries of water, formaldehyde (H_2CO), carbon monoxide (CO), molecular hydrogen (H_2, found by its ultraviolet lines), hydrocyanic acid (hydrogen cyanide, HCN), more diatomics (carbon monosulfide, CS; silicon monoxide, SiO; and sulfur monoxide, SO), dimethyl ether (CH_3CH_3O), the ever popular ethyl alcohol, and the not so popular methyl version (CH_3OH).

Ninety interstellar molecules and molecular ions are now known; another dozen or more are found only in circumstellar environments, in gaseous envelopes surrounding stars. Quite a few are simple ions (for example, the formyl ion, HCO^+; the sulfoxide ion, SO^+; the hydronium ion, H_3O^+; and the familiar CH^+). A few others, like HCN (hydrogen cyanide) and HNC (hydrogen isocyanide), or dimethyl ether and ethyl alcohol, are isomers, structural variations of each another. Many molecules have numerous observed lines, and with modern sensitive radio receivers we see that the microwave spectrum is loaded with them.

Interstellar chemistry has truly arrived, the result of a powerful amalgam of electrical and radio engineering, astronomy, chemical theory, and chemical-laboratory spectroscopy that can determine the frequencies of the molecular lines. Included in the list of interstellar discoveries, however, are several molecules and radicals that are not seen—and cannot be created—in terrestrial laboratories. Quantum theorists therefore had to calculate approximate frequencies. The radio astronomers could then determine exact frequencies (which their lab counterparts could not do) and moreover find other lines, the data helping to establish the molecular structures. The exercise is a fine example of the synergy among the sciences, a demonstration that there are not "sciences" so much as "science." It also shows the fallacy of categorizing astronomy (and other sciences) as "pure sciences," intellectual disciplines with no practical applications.

Included in the set of interstellar molecules are fascinating nitrogenated carbon chains with uneven numbers of carbon atoms. They begin with stable HCN and HC_3N (cyanoacetylene, structurally HCCCN) and then extend to unstable, and unearthly, HC_5N through HC_9N—cyanooctatetrayne, an 11-atom molecule! Only under conditions of low temperature and density can such species survive. Though ring-shaped hydrocarbon molecules (C_3H_2, cyclopropenylidene, for example) are probably more abundant, fewer such varieties are seen because their radiation is spread over more lines, resulting in weak line strengths. This contrast illustrates one of the most severe problems in astronomy: observational selection. We observe the things most likely to be seen, not the most abundant, and must at times make theoretical allowances for the unobserved. In stellar astronomy, selection makes the naked-eye sky seem filled with hot, bright B and A stars, whereas by far the most common are the faint M dwarfs, none of which can be seen without optical aid.

Another unusual collection of molecules with a common property is the "protonated" molecule, a molecule to which a proton—a hydrogen ion—is added to produce a molecular ion: examples are HCO^+ and N_2H^+. Although these are quite unstable, they can exist for long periods under low-density conditions. Yet another form is the "deuterated" molecule, in which the heavy isotope of hydrogen, deuterium (^2H), replaces one or more of the ordinary ^1H atoms, creating such examples as ^2HCN (or DCN). The increased mass of the heavier isotope changes the vibrational and rotational characteristics of the molecule and significantly shifts the frequency of the line, allowing the deuterated molecule to be separated from the normal version. (The same phenomenon allows stellar astronomers to derive isotope ratios from molecular lines in cool stars.) Such isotope substitutions are seen for other kinds of molecules as well (for example, ^{13}CO is easily separated from the more ordinary ^{12}CO), allowing isotope ratios to be compared with stellar or terrestrial ratios.

Isotopic variations, while not of great significance in terrestrial chemistry, have great astronomical importance: different isotopes are manufactured in stars as a result of a huge variety of nuclear reactions and are then launched into interstellar space aboard powerful stellar winds (outflows of mass akin to the solar wind) and the detritus of supernovae, exploding stars. Examination of interstellar isotopes thus gives us a way of examining the history of the interstellar medium and even of checking theories of stellar evolution and death.

Perhaps the most intriguing interstellar molecule of all, one that might herald a remarkable group, is interstellar glycine (NH_2CH_2COOH), tentatively identified in 1994. While not the most complicated molecule, it is certainly one of the most sought, for it is the simplest of the amino acids, the foundation stones of proteins, in turn the foundation stones of life. Support for its existence was the firm discovery in 1996 of CH_3COOH, acetic acid—vinegar—which is expected to combine with ammonia to form glycine. The 1970s were the dawn of astrochemistry; will the '90s be seen as the true dawn of astrobiology?

Solid Stuff

Atomic and molecular interstellar lines, those that arise from the gaseous state, are sharp, narrow features, whether seen in emission or absorption. Even the 21-cm line is narrow when we look at individual clouds and allow for Doppler effects along the line of sight. However, in the 1920s astronomers began finding a number of broad, shallow features in the spectra of distant stars. These were

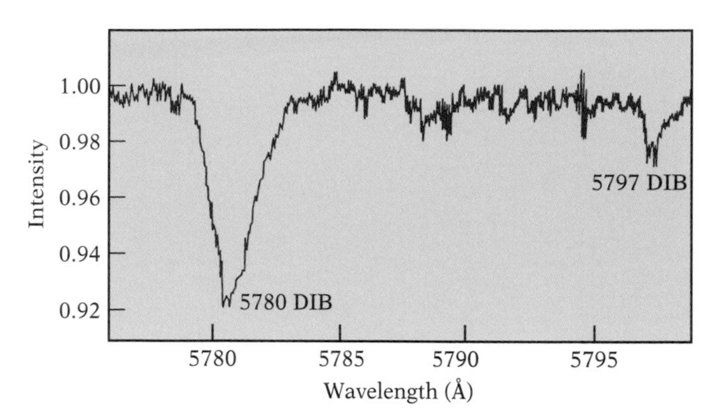

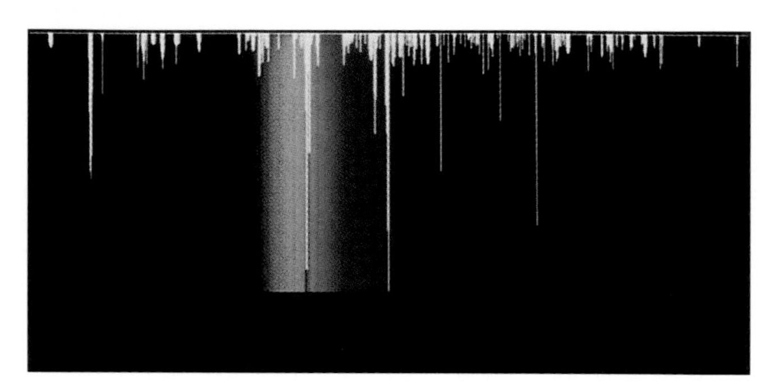

The diffuse interstellar bands (DIBs), broad unidentified depressions in stellar spectra (above), are seen from the ultraviolet through the near-infrared (at right).

clearly interstellar in nature, as their strengths correlated with the degrees of interstellar extinction suffered by the background stars. We know of over 100 of these "diffuse interstellar bands," or "DIBs," that cover the spectrum from the ultraviolet through the near infrared, to wavelengths as long as 13,200 Å (1.32 μm); the best known and most prominent is at 4430 Å.

None of the DIBs has yet been identified, though there are several candidates. Their broad, washed-out appearance suggests that they might be formed by absorption on some kind of solid surface, where interactions between closely spaced atoms strongly widen energy levels. One fascinating possibility is that they are produced by very large carbon-bearing molecules. They may be created by "polycyclic aromatic hydrocarbons," or "PAHs," molecules composed of linked benzene rings (hence "polycyclic") and called aromatic because of their strong odors. PAHs may grow so large as to be considered small dust grains. Curiously, benzene itself is not observed, probably for the same reason other ring hydrocarbons are absent: the lines are just too weak. PAHs are also believed to be responsible for a set of emissions observed in the infrared between 3.3 μm and 11 μm against both diffuse and reflection nebulae. Another, less likely, possibility for the emissions is the "bucky ball," buckminsterfullerene (C_{60}), named for the architect who designed the geodesic dome, which the structure of the molecule resembles. Variations on these structures might also produce the DIBs, as could a variety of other compounds, including (according to a recent study) simple H_2.

The DIBs and the large molecules are a bridge to the solid state. Just as free molecules in the gaseous state can rotate and vibrate, chemical bonds in solids can stretch and bend with quantized energy levels. In the infrared we see absorption features at 3.08, 13.5, and 45 μm firmly linked to the laboratory spectrum of ice (just right to chill the interstellar ethanol). The interstellar line is not exactly at

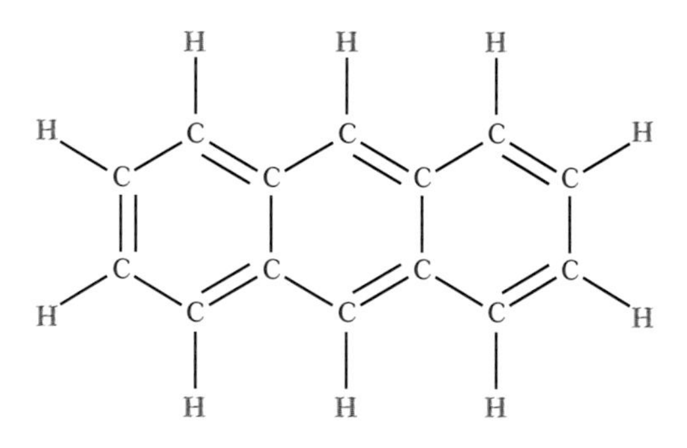

Benzene rings can link into a large number of different kinds of polycyclic aromatic hydrocarbons (PAHs)—here, anthracene.

the lab wavelength, probably because the ice is impure and is not crystalline but amorphous—simple solid water of a kind unknown in Earthly refrigerators. Other confirmed solids are carbon monoxide and, surprisingly, methanol, CH_3OH, which is also seen in the gaseous phase. Dry ice (frozen CO_2) is a possibility; more uncertain are methane (CH_4), hydrogen sulfide (H_2S, adding to the smell of the PAHs), and some other carbon–nitrogen–oxygen compounds. These volatile solids are almost certainly mantles deposited on and coating interstellar grains.

We therefore come back to the grains themselves, the dusty stuff that dims starlight and is responsible for globules and the dark constellations of the Incas. The 2200-Å bump in the ultraviolet seems to be caused by amorphous carbon, perhaps graphite, a distant cousin to the bucky balls. We also see evidence for an absorption caused by the quantum stretching of a CH bond that could be produced by hydrogenated amorphous carbon, a cousin of the PAHs. And, finally, we return to a 9.7-μm SiO band that results from bond stretching and an even "redder" bond-bending band that reveals the presence of silicates in interstellar space. The depletion of gas-phase metals suggests interstellar metal-rich silicates, perhaps even in the form of olivine, a mineral found in abundance on Earth.

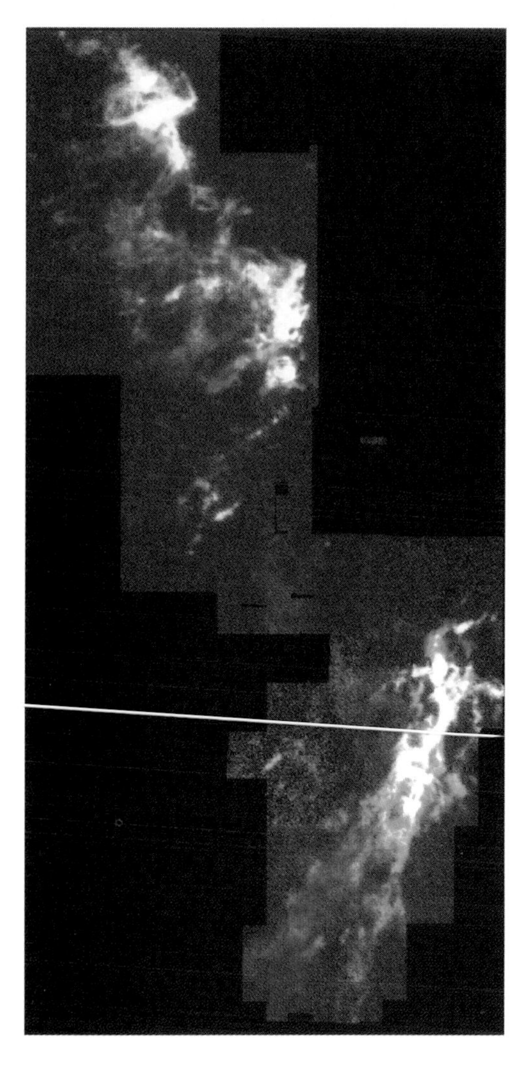

Molecular clouds in Orion, traced here by emission from ^{13}CO, are associated with the dark clouds Lynds 1641 (the lane toward lower right) and Lynds 1630 (that toward the top). The Orion Nebula lies near the top of the lower bright lane, the Horsehead Nebula near the bottom of the upper lane. Infrared radiation from the clouds' dust can be seen in the lower half of the image on page 77.

Molecular Clouds

Interstellar space is filled with ultraviolet photons from hot stars, as is obvious from the existence of the low-density, warm ionized medium and the faint background of Hα-emitting gas seen all over the Galaxy. Such high-energy radiation is death to molecules because it easily splits the fragile bonds between atoms. We therefore find no molecules within the warm ionized medium, nor even within the warm neutral medium. Even the denser H I clouds are too tenuous to produce—and protect—much in the way of molecules.

The gas within all the various kinds of dark clouds, from the huge complexes and rifts to the tiny black globules, is shielded by dust from the damaging radiation. Lacking one of the principal heating sources of interstellar space, the gas cools to low temperatures, in some cases only a few kelvins above absolute zero. Welcome now to the final portion of the interstellar medium, to the missing 25 percent—the *cold* component, 15 to 20 K, as measured from the ratios of molecular emission lines.

The globules, which are rich in molecules like CO, claim the right to be called molecular clouds, albeit small ones. Larger ones, "giant molecular clouds," or "GMCs," are everywhere, but not always as obvious. All the complex molecules of interstellar space are found in giant molecular clouds. Some GMCs are associated with optical features like the great stellar voids of Taurus and Auriga and those in Ophiuchus that were first recognized in the early twentieth century by E. E. Barnard. More subtle are GMCs like the one in southern Orion connected with the Orion Nebula. We know it is there only because of intense emission of molecular gas, in particular CO. Many GMCs are hidden behind foreground dust and can be viewed only in the radio spectrum, the outstanding example the Sagittarius B2 cloud, the richest known source of interstellar molecules, containing almost all those ever found. There and Orion are the first places astronomers look.

The major constituent of stars is hydrogen, and the globules and giant molecular clouds from which the stars come must be hydrogen dominated too. Ultraviolet shielding and bone-chilling temperatures, however, put hydrogen into its molecular state, H_2. Molecular hydrogen is very difficult to observe because it is homonuclear—that is, it has identical atoms—and consequently has no permitted rotation–vibration bands and no radio lines. It can be detected in absorption and emission by ultraviolet electronic bands, but only if starlight penetrates the gas, precluding its general observation in dense molecule-bearing clouds. There are also forbidden rotation–vibration bands in the infrared, but they are weak and difficult to observe. But where we can make the measurements, we find that H_2 dominates, showing that in the cold clouds essentially all the hydrogen is indeed in its molecular form.

The amount of H_2, as measured from its column density, is usually estimated by using an easily observed tracer that is expected to be in constant ratio to H_2; the most common is CO, which has intense emission lines. The problem is to find the "*X* factor," the ratio of the column density of H_2 to the intensity of ^{12}CO emission integrated along the line of sight. The amount of the elusive H_2 is difficult to measure and is subject to considerable error. It can be estimated directly from its weak lines as well as by indirect methods. Assuming that most of the gas of a selected cloud is in the form of H_2, we can measure the mass of dust from its infrared radiation and assume a likely gas-to-dust ratio; or, we can estimate the density of the gas from the rate at which it collisionally excites other molecular emission lines. The problem then is to relate these cloud-mass or cloud-density measures to CO column densities. Although subject to considerable controversy, the *X* factor is still well enough known to be useful in estimating cloud masses. The CO radiation allows easy mapping of the clouds and the measurement of angular dimension.

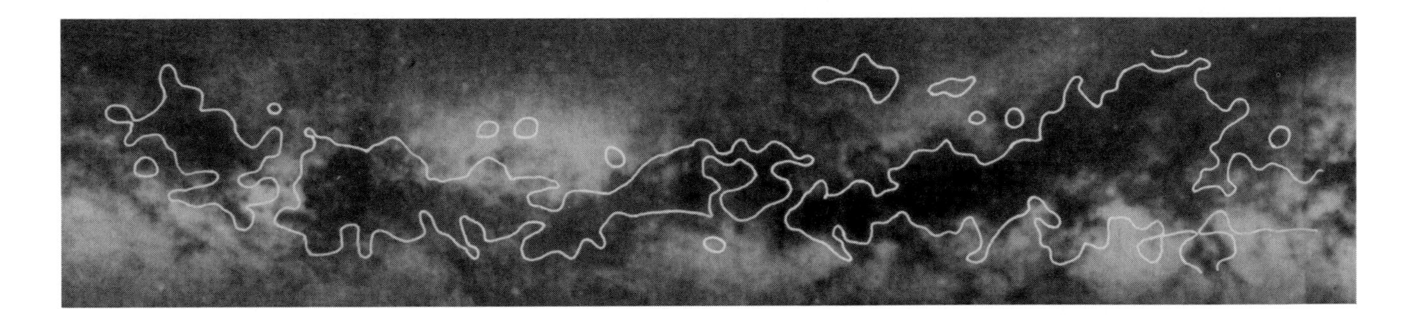

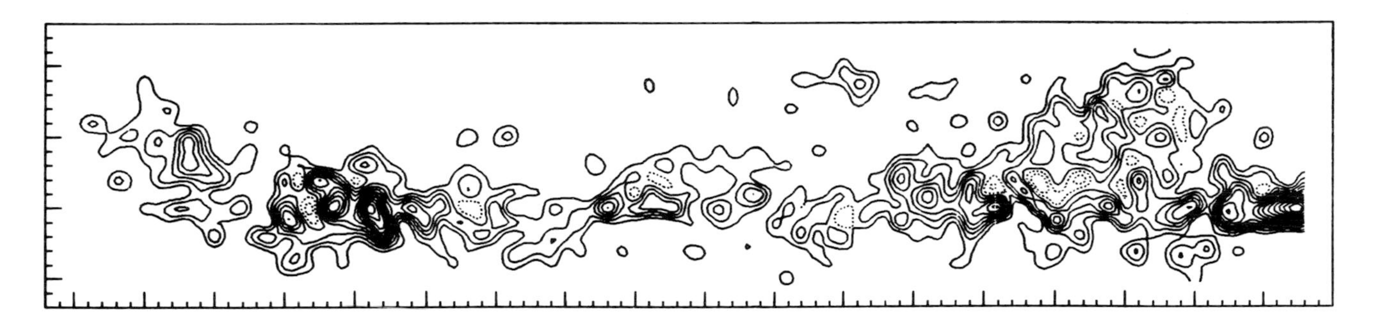

Giant molecular clouds, outlined by CO emission (bottom), throng the Milky Way. The clouds clearly lie along the dark, dusty, cold rifts (outlined in the photograph above).

Distances can be estimated by using the galactic rotation curve and the radial velocities given by the emission lines or derived from related stars, yielding physical dimensions, masses, and average densities. Masses can also be estimated by measuring internal velocities through Doppler shifts and determining the mass needed to hold the cloud together under its own gravity.

The cold clouds have an enormous range of properties. The smaller clouds and globules are a few parsecs across, with masses of perhaps 10 to 1000 solar masses and densities of 100 to 1000 molecules cm^{-3}; complexes of the smaller clouds can have masses of 10^5 solar masses. The giants, the GMCs, extend to low tens of parsecs across with masses of thousands to 10^5 solar masses; complexes of GMCs can be 100 pc across and contain over a million solar masses. Densities are in the hundreds and thousands of molecules cm^{-3}, but the clouds are very inhomogeneous and have embedded clumps with densities that range into the hundreds of millions. The GMCs are the most massive objects in the Galaxy. Over 2000 are known, spread throughout the Milky Way, and together they make up the missing quarter of the known interstellar medium's mass.

Like H II regions and H I clouds, GMCs outline spiral arms. Otherwise, the difference between the distribution of molecular and neutral atomic hydrogen is striking. Except for a "hole" surrounding

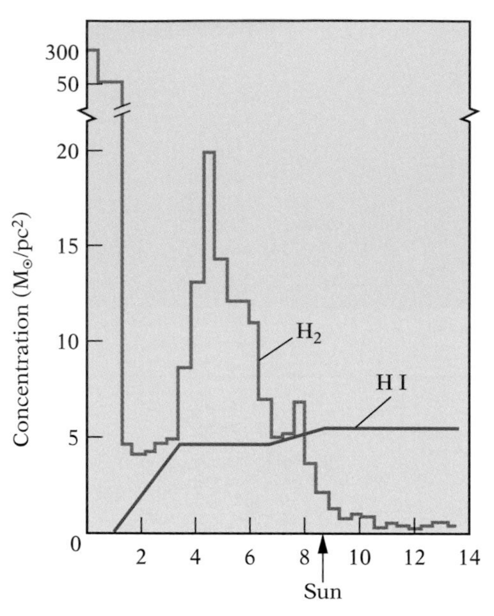

Like neutral hydrogen (H I), giant molecular clouds fall along spiral arms. Unlike H I, however, molecular hydrogen (H_2) concentrates to the galactic center and into a ring with a radius of about 5 kpc.

the galactic center, H I is spread rather uniformly throughout the Galaxy out to beyond 15 kpc. Molecular hydrogen, however, is concentrated strongly to the center, where it seems to be in a state of expansion, and to a ring—perhaps a massive spiral arm—between 3 and 7 kpc that contains some 2 billion solar masses. The H_2 abundance then drops sharply beyond the solar circle. Some of the difference is the result of an increase in the pressure of the interstellar medium inward, toward the center of the Galaxy. Higher pressure means a higher density and a greater degree of shielding of the clouds against the encroachment of high-energy stellar radiation, and thus a greater degree of molecule formation.

Brick by Brick

The first clue to the construction of interstellar molecules from the atoms of interstellar space lies in their relative abundances. Astronomers derive the chemical compositions of the clouds—the ratios of different molecules to H_2—from the measured column densities. Uncertainties arise, however, because of the *X*-factor problem, difficulties in assessing optical depths, occasional poor knowledge of absorption or emission probabilities, and the clouds' severe inhomogeneities, their variations in density along the lines of sight.

After H_2 itself, CO is the most abundant observed variety, with roughly 1 CO molecule for every 10^4 or 10^5 of H_2. Generally, the more complex the species the less there is of it. Methyl alcohol is down from molecular hydrogen by a factor of perhaps 10^8, and the long chain structures are decreased by maybe 10^{13}—it is a wonder that these molecules can be detected at all. The results are highly incomplete, as several molecules that should be abundant, like O_2, are not observed. Moreover, there is not just *one* set of abundances. Each cloud observed is different, and since astronomers tend to observe the clouds in which most of the molecules are found—the richest ones—what is found may not even be typical, in another example of observational selection.

The great theoretical challenge is to understand how these molecules are formed in the low-density, low-temperature environment of interstellar space. At best, we wish to reproduce the abundances from known processes through reaction networks that have thousands of steps and must include molecules that are not observed. The abundance of a particular species is the result of a balance between creation and destruction, including the reactions that make one kind of molecule from another. Interstellar molecules can form easily only if the processes are exothermic—that is, if they release energy—as there is insufficient driving energy in the cold clouds for endothermic, or energy-absorbing, processes. The processes must therefore involve the emission of photons or the deposition of energy on the ubiquitous grains. The most rapid gas-phase reactions involve ions, so there must be a way of creating them.

Interstellar chemistry starts with the dominant molecule, H_2. In the gaseous state it can be formed by the attachment of a free electron to a hydrogen atom to create H^- and a photon, the reaction followed by the collision with a neutral atom to make H_2 (which rejects the third electron). It may be surprising that a hydrogen nucleus, a proton, can stably support two electrons. In the Bohr atom, if you—an electron—are close to a proton with the attached electron on the far side, you experience a positive residual electric charge that can loosely bind you. The H^- ion is the principal source of the great opacity of the solar atmosphere and the reason the gaseous Sun seems to have a sharp edge: the second electron is easily broken away by absorbing light. Molecular hydrogen can also be formed by the collision of a positive ion—a proton—and neutral hydrogen to make H_2^+ and a photon, this process followed by a collision with a neutral atom that gives up its electron.

However, under low-temperature conditions gas-phase H–H reactions are very slow and will not suffice to make the extraordinary amounts of H_2 observed. Instead, the molecules have to form on grain surfaces when a hydrogen atom collides with a grain to which hydrogen atoms are already attached. The resulting release of energy can be absorbed by the grain, and that energy (together with grain–grain collisions) can kick the newly formed H_2 molecule back into space, into the gas phase.

From here, the long process is reasonably well understood, at least in its elementary form. There is no chemical equilibrium of the kind that works in a terrestrial laboratory. Instead, the reactions occur one at a time, brick by brick, building heavier and heavier compounds. They start with the ionization of H_2. To exist at all, however, the molecules must be shielded from destruction by ultraviolet radiation by the same grains that form them. Therefore, no photons are available for ionization either. How, then, can the process begin?

There is an alternative to photoionization. Interstellar space is full of cosmic rays, atomic nuclei that fly near the speed of light. They appear to be created by the energetic results of supernova explosions. Cosmic rays continuously crash into the Earth's atmosphere, where they break up atoms and produce showers of secondary particles easily detected on the ground. They are a bane of the optical observer, as they can create false images on CCDs.

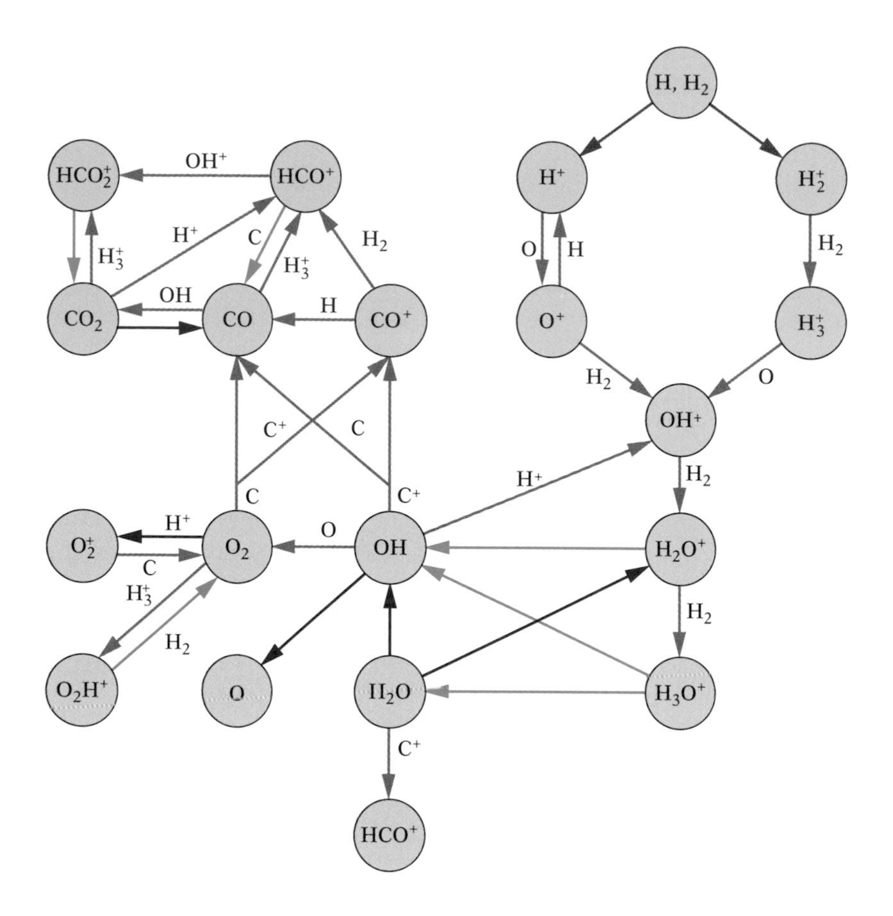

This theoretical reaction network shows the development of oxygen-bearing molecules: reactions aided by cosmic rays are shown in purple, by electrons in green, and by photons in orange.

Sleeping astronauts have even been awakened by the cosmic rays hitting their retinas, producing apparent bursts of light. Though hydrogen nuclei are the most common cosmic rays, nuclei heavier than iron have been seen. The energies can be spectacular, the greatest observed close to that of a professionally pitched baseball, from *one nucleus* traveling close to the speed of light.

With their fierce energies, cosmic rays can penetrate dark, dusty clouds that photons cannot and can produce the necessary ionization of molecular hydrogen. Ionized molecular hydrogen subsequently is hit by a neutral H_2 molecule to make (unobserved) ionized "trionium," H_3^+, releasing a neutral atom in the process. Trionium and a neutral oxygen atom then collide to make OH^+ and H_2. The ionized hydroxyl then *picks up* H_2 to make ionized water (plus H), which in turn picks up yet another H_2 molecule to create weird H_3O^+ (plus H). The ionized water and H_3O^+ collide with free electrons to make neutral water and OH (plus by-products). This chain at least gives the flavor of the process; numerous other reactions with oxygen, hydrogen, and carbon follow that can build CO, HCO^+, CO_2, and other molecules. Other reaction networks that are thought to take place on grain surfaces within dense clumps (like the Heimat source of Sagittarius B2 North) appear to produce the large hydrogenated organic molecules.

Remarkably, theoreticians have been able to come close to the observed compositions, showing that the basic idea of ion-driven reactions is correct. Yet there are some equally remarkable discrepancies, with the abundances of molecules like C_3H_2 and HC_5N and others off by several factors of 10. We are really only at the beginning of understanding a *very* complex set of different reaction systems that can make each cloud unique. Theoreticians must consider a variety of other factors that include high-density clumps that can produce severe place-to-place composition variations, photon heating by embedded stars, shock heating from winds from massive stars and supernovae, chemical reactions that continue on grain surfaces, depletion of atoms onto grains, and large molecules—many, perhaps *most* of them, not yet identified—that can alter the reactions.

The ultimate goal is to string the different molecular clouds into an evolutionary sequence. Their compositions must continually change as the clouds evolve, as they compress under the force of gravity, as they collide, as they are hit by energetic photons, as winds from hot stars dissipate them. We may then eventually be able to use the observed compositions to tell the states of the different dark molecular clouds and see where they fit into the general flow of cloudy life.

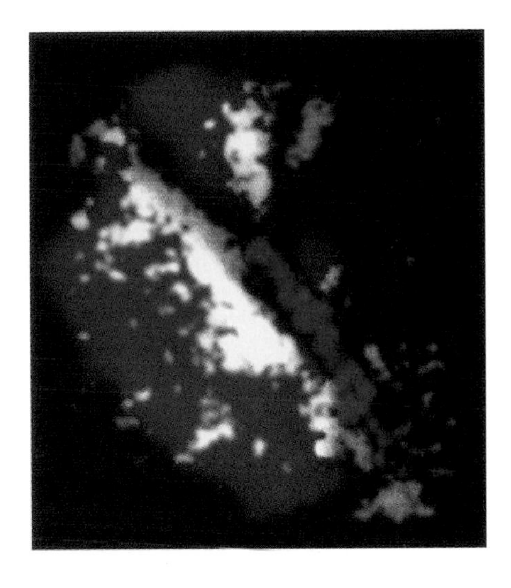

The bright bar of the Orion Nebula, seen in the optical on page 28, is expanded here in three wavelength regions to show the transition between the ionized and molecular clouds. (The ionized zone itself, not shown, is out of the frame to the upper right.) At the outer edge of the ionized zone lie excited PAHs (blue) as revealed through the infrared 3.3-μm emission line; in the middle and farther into the dust is a band of ultraviolet-emitting excited H_2 (green), and farther yet into the dust CO is emitting in the radio (red). To the lower left the dusty molecular cloud is filled with H_2, CO, and all kinds of other good molecular stuff.

WARFARE

Now expand the view to include the total interstellar medium. None of the different kinds of clouds—H I, H II, molecular—exists in isolation from the others. Indeed, all are engaged in an ongoing struggle for ascendancy, as clouds collide, as high-pressure clouds push against the low-pressure intercloud media and these push back, and as shock waves from supernovae are plowing through the whole affair. Cloud interactions are best seen at the boundaries where bright matter clashes against dark. The most obvious example is the classic Strömgren sphere, in which ionizing photons from hot exciting stars create a diffuse nebula within a neutral medium—look again at the Rosette Nebula in Chapter 2 (page 53).

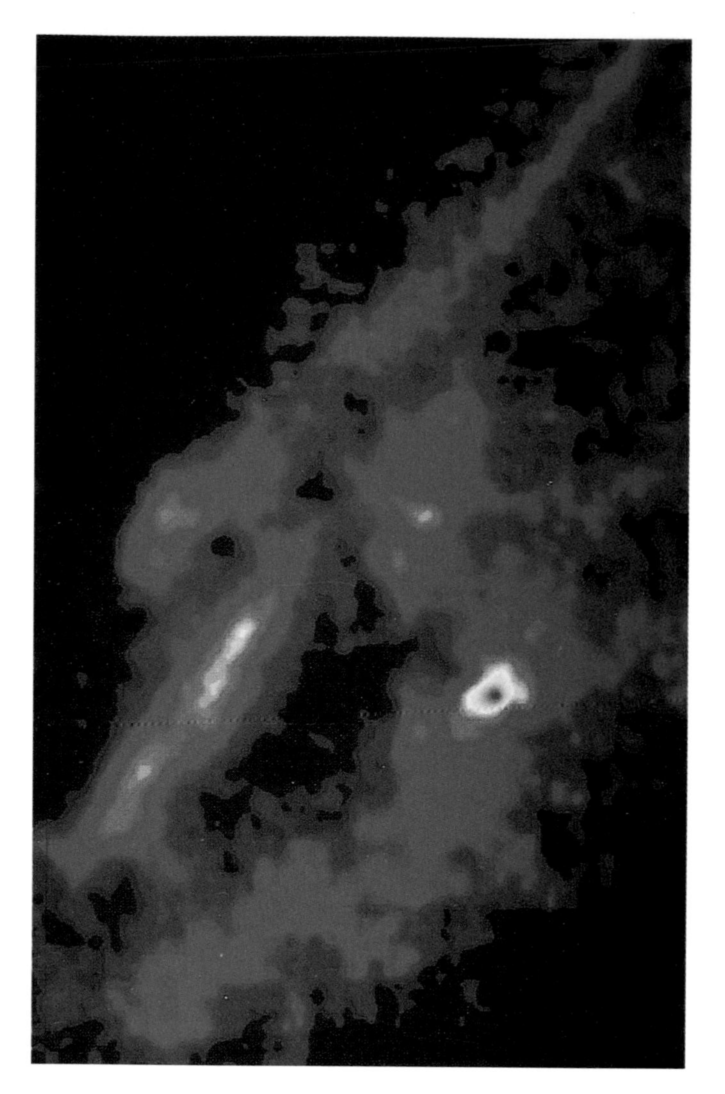

On the left, in a field of view about 1.5 degrees long, is $C^{18}O$ emission from the nearby Taurus molecular clouds, hotbeds of star formation that we will shortly encounter. On the right is the same field seen in the optical, the cloud lanes made visible by thick dust and the absence of stars.

With the aid of infrared and radio telescopes, we now know that the neutral cloud surrounding such a bubble is not simple neutral hydrogen but hydrogen in its molecular form, that the diffuse nebulae are commonly bubbles within, or blisters at the edges of, molecular clouds. The interaction, wonderfully complex, is illustrated beautifully by the Orion Nebula. The bright Orion "bar" is an ionization front, a distorted piece of a Strömgren sphere seen on edge. The visible bar, however, is but the vanguard of a smooth transition zone into the dark cloud. High-energy hydrogen-ionizing photons beyond the Lyman limit are finally stopped at the optically visible zone. But lower-energy ultraviolet radiation, that not capable of ionizing hydrogen, can penetrate beyond it into the dust, where it can ionize carbon and other atoms and also excite and break up molecules. Such "photodissociation regions," or "PDRs," are among the hottest current topics in molecular astronomy, as they reveal more and more about how different kinds of clouds develop and evolve.

A PDR is the molecular analogue to the Strömgren sphere, a cloud region in which molecular dissociation is in balance with molecular formation, just as within the diffuse nebula hydrogen ionization is in balance with recombination. Thus in a PDR we have additional possibilities for astronomical chemistry. The structure of a PDR is controlled by the ability of the photons to penetrate the ever-present dust. At the inner edge, where dust extinction is not too great, we find excited and hardy PAIIs; farther into the darkness, neutral atomic hydrogen is converting to molecular hydrogen, and, farther yet, carbon converts to CO. The PDRs allow us to come full circle in both time and space, firmly joining the obvious H II regions, first studied 200 years ago, with the molecular clouds revealed by modern advances.

By so linking different aspects of the interstellar medium, by examining their relationships and evolution, we begin—brick by brick—to reconstruct the history of our own development from the vastness of interstellar space.

6

STAR FORMATION

≺ *The Hubble Space Telescope reveals dramatic details of elephant trunk structures light-years in length within the Eagle Nebula. Their tips, evaporating under radiation from hot stars above the picture, contain new stars that were created within.*

With great deliberation, Black God of the Navaho reached out from the hogan of creation to place the bright stars and the constellations into the sky. The great troublemaker Coyote, miffed at not being consulted, stole Black God's pouch of crystals and scattered the remainder of them over the face of the heavens. Now, night after night, the stars wheel around us, "forever," wrote Addison, "singing as they shine."

What really happened? Have the stars really existed "forever"? If not, they must have birthplaces and some celestial mechanism must have given them up to the heavens. And if that is so, maybe stars are being made today. Perhaps by watching stars forming now we can see the birth of our Sun mirrored in Black God's dark sky.

First Clues to Formation

The initial evidence for the means of stellar creation is found within our own Solar System. The dominant Sun, an ordinary yellowish G2 dwarf with a surface temperature of 6000 K and a mass of 4×10^{33} grams, is surrounded by a wild variety of debris, chief among them the planets. Four small bodies—Mercury, Venus, Earth, Mars—orbit within 1.5 AU of the Sun; four larger ones—Jupiter, Saturn, Uranus, Neptune—revolve much farther out, at distances of 5 to 30 AU. Even the largest planet, Jupiter, has a mass only 1/1000 that of the Sun, and our Earth (the biggest of the inner four) weighs in at a mere 1/300,000 solar.

The Sun's family is characterized by a comprehensive order. The planets all orbit the Sun in the same direction—counterclockwise looking down from the north—and are tightly confined to the same orbital plane. Only the innermost (Mercury) and outermost (Pluto) deviate much, having orbital tilts of 7° and 17° respectively to the plane of the Earth's orbit (and, as we will see, there is serious doubt about calling Pluto a planet). All except Venus, Uranus, and Pluto

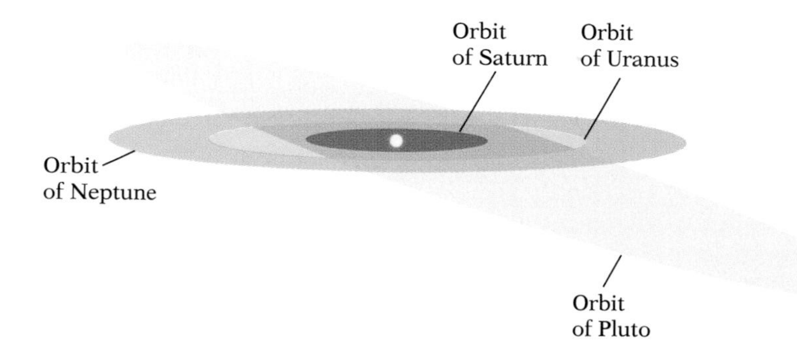

The planets orbit the Sun on nearly circular solar orbits in nearly the same plane, defining a disk about the Sun.

spin in the same direction in which they orbit. Most significant, the planets orbit and spin in the same direction in which the Sun rotates, their orbital planes lining up closely with the solar equator. We are therefore inevitably led to the belief that the planets are byproducts of the Sun's formation.

This conclusion is nearly 250 years old. Before Immanuel Kant published his *Critique of Pure Reason* in 1781, he had explored several questions of science, including the cause of earthquakes, the tidal slowing of the Earth, and the formation of the Sun and planets. In 1755, he suggested that the Sun was born as the central core of a vast gaseous mass that, upon contracting under the force of gravity, spun itself into a disk that condensed into the planets. Fifty years later, the French mathematician and astronomer Pierre-Simon de Laplace assured readers of his *Traité de mécanique céleste* of the long-term stability of planetary orbits. Unaware of Kant's brilliant deduction, he reintroduced the "nebular hypothesis" in a footnote to his popular discussion *Exposition du système du monde.* The Sun and planets were apparently born together from a spinning, flattened disk, a "solar nebula." To understand the birth of the Sun, and by analogy the stars, we must therefore look to the Solar System as a whole.

Asteroids, debris concentrated between the orbits of Mars and Jupiter, still fall to Earth as meteorites. The oldest things known, the meteorites date from the time of the formation of the Solar System.

The ordered planets contain some 98 percent of the mass of the Sun's family. The remainder provides additional important clues to formation. Thronging the space separating Mars and Jupiter are countless bodies that range from pebbles to massive mountains 1000 km in diameter: these arc the asteroids, the remains of a failed planet. A few make it into the inner Solar System, a small subset screaming through the atmosphere as meteors, some crashing to Earth to become meteorites.

The discovery of radioactivity by Marie Curie in the early twentieth century allows measurement of the meteorites' ages. All the isotopes of the Earth's heaviest natural element, 92-proton uranium, are unstable. Atoms of ^{238}U, by far the most common isotope, spit helium nuclei and gamma rays, decaying down the periodic table to create in sequence a variety of elements from protactinium (atomic number 91) through radium (atomic number 88) and stopping finally at stable lead (^{206}Pb, atomic number 82). The decay rate is steady and predictable: if you start with a kilogram of ^{238}U, in 4.5 billion years you will have half a kilogram of it left, in 9 billion a quarter of a kilogram. Many rocks contain small amounts of uranium, and thus the ratio $^{206}Pb/^{238}U$ gives a rock's age, the time since its solidification. (The original fraction of ^{206}Pb is known from rocks that do not contain uranium.) The ages of Earth rocks range from zero (for cooling lava) to an astonishing 4.2 billion years. Rocks brought back from the Moon date to nearly 4.5 billion years,

Comet Hyakutake sports a tail of ionized gas and dust 40 million km long created by the action of sunlight on a dirty iceball only 2 km in diameter.

and the oldest rocks known, the meteorites, to 4.6 billion. The Solar System must be at least as old as its oldest rocks. The similarity in the ages of the Earth, the Moon, and the meteorites, combined with the meteorites' planetary nature, implies that all the components of the Solar System are the same age, 4.6 billion years, and that any differences are due to specific formation times. Kant's spinning disk is indeed ancient, and Laplace was right about the stability of the system.

Comets allow investigation of the Solar System beyond the planetary orbits. They too revolve around the Sun, but on long elliptical paths that can take them far beyond the confines of the planets. The most famous comet, Halley's, invades the orbit of Earth every 76 years and turns around near the orbit of Pluto. Unlike asteroids, comets are loaded with ices (of water and other chemicals) that sublime away as the bodies approach the heat of the Sun. The gases ionize under sunlight and are then pushed away by the solar wind and wrapped by the solar magnetic field to create bright ion tails. Dust and fragile rocks embedded within the ices are subsequently released and pushed away by sunlight to create dust tails illuminated by the Sun's reflection. Some of the solid debris hits the Earth's atmosphere to appear as meteoric streaks slicing the sky.

Comets are fragile structures easily destroyed by solar heat (Halley's will be gone in 10,000 years or so), but new ones are continuously being found. There must be a huge reservoir of them, and in fact there are two. Comets are differentiated by their orbits, which can be determined by observing the comets' positions when they are near the Earth. The long-period comets, those with periods of more than 200 years, have random orbits not fixed to the Solar System's plane. Most of them have periods longer than 100,000 years, so they must come from very far away. Yet all have closed elliptical paths, so they must still belong to the Solar System. Their reservoir is the great Oort comet cloud, named for the Dutch astronomer Jan Oort, who first hypothesized it. The cloud extends a good part of the way to the nearest star, and most of its comets will never get near the Earth, but a few are disturbed by the gravity of passing stars and interstellar clouds and come plummeting toward us. The short-period comets tend to be confined to the Solar System's general plane and orbital direction. Their home is the Kuiper belt, identified by the Dutch-American astronomer Gerard Kuiper, which extends the planetary disk to great distances from the Sun.

A logical pathway from present-day evidence takes us backward in time. Look at the ancient face of the Moon, cratered to the point

The Moon is covered with craters on top of craters, holes punched in the crust by meteorites. The giant dark areas are meteorite basins overlaid with extruded lava.

of destruction by immense meteoric bombardment. Comets and asteroids still strike the Moon (and Earth); but they must have been vastly more numerous in the past, indicating that they are "planetesimals," the bodies that assembled to make the planets. The comets are, as shown by their tails and by meteors, made of ice and dust. Ice is frozen gas. The comets and planets were therefore assembled from a dusty, gaseous disk. The Kuiper belt comets look like remnants of the original planetesimals that formed the outer planets; the rockier asteroids (also in the disk) were the precursors of the inner planets, those like the Earth.

The structure of the Solar System gives us some idea of what we are looking for. Can we look out into the cosmos to catch stars and systems like ours in the act of development? Can we use our observations to reveal what happened here 4.6 billion years ago? Just watch.

Youth

The first step is to identify young stars, as their locations should readily reveal the sites of star formation. Hot O and B stars, only recently escaped from their weakly bound associations, fit the bill. Their intimate alliance with the diffuse nebulae they excite strongly suggests that these lovely gas clouds are the remnants of star birth. Even Herschel could see the relations between the bright and the dark clouds, and thus we return to Russell's suggestion that the globules are stellar nurseries. If the dark molecular clouds produce hot stars, we might expect them to deliver cooler ones like our Sun as well. Since O stars and molecular clouds define the Galaxy's spiral arms, the arms must also be related to star formation, the luminous stars acting as markers of stellar birthplaces in our Galaxy and in others.

To locate the lower-mass young stars we need look only a little deeper. Clumped in and around dark cloud complexes lurk packs of strange "variable stars," stars that change their brightnesses with time. Dozens of different kinds of variables are known, most of them stars of advanced stellar age. Different classes of variables are commonly named after the prototype, the first one of a kind to be found. The T Tauri stars (Roman letters commonly identify variables), named after the star T Tauri, which is located in front of a dark cloud in Taurus, were first identified as a group by A. H. Joy of Mount Wilson Observatory in 1945. Over the years since their discovery, they have given us powerful clues not of stellar death but of birth.

Most T Tauri stars—hundreds are known—fall into spectral classes F, G, K, and even M, their temperatures thus ranging between about 7500 K and 3000 K. Erratically variable, they change their brightnesses by a few tenths of a magnitude up to a few magnitudes over intervals of days and months. Their spectra contain strong emission lines of hydrogen (Hα, Hβ, etc.); forbidden oxygen and sulfur ([O I] and [S II]) lines commonly appear as well. Such emission lines are produced by a hot gas under low pressure, as in a diffuse nebula, and therefore indicate circumstellar clouds of gas. T Tauri stars are also associated with reflection nebulae and have infrared continua that are anomalously strong—stronger than expected for a blackbody at the stars' temperatures—and that are most likely produced by heated circumstellar dust. All these characteristics, especially when taken together, tie the stars even more firmly to the surrounding interstellar medium.

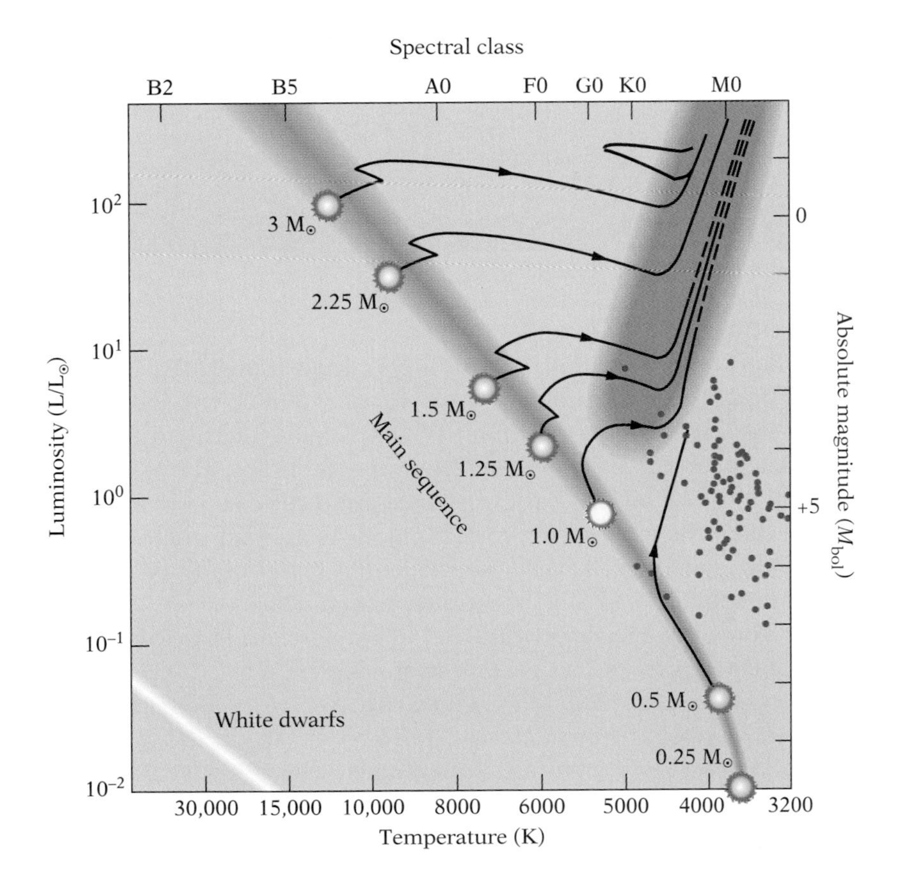

Giant stars (their locus the shaded pink area) lie above the main sequence and along evolutionary tracks computed for aging solar-type stars that are exhausting their internal hydrogen. Bolometric magnitude —M_{bol}—is absolute magnitude corrected for radiation outside the visible spectrum. T Tauri stars (blue dots) also lie above the main sequence, but outside the bounds allowed by the evolutionary tracks. Their youth implies that they are moving toward *the main sequence.*

This optical view of NGC 2024, just above Zeta Orionis in Orion's belt, shows a dark lane cutting the diffuse nebula in half. A look in the infrared reveals hundreds of embedded stars, which must have been born recently in the clouds.

T Tauri stars gang into expanding T associations that, like OB associations, are indicators of youth. Equally telling, the stars have strong absorption lines of lithium, a fragile element easily destroyed at high temperatures by nuclear reactions. Stars like the Sun have convecting outer layers that cycle their atmospheres deep into hot layers where the reactions can take place. As a result, the Sun and other older stars have little of the element. A high lithium content is therefore another indicator of extreme immaturity.

We know the distances of the T Tauri stars from those of the associated clouds, and thus the stars' luminosities. When T Tauri stars are plotted on the HR diagram, they are seen to be brighter than expected for their spectral classes, falling well above the main sequence. But so do giants and supergiants. The different kinds of stars, which lie in different parts of the HR diagram, are connected theoretically through the construction of evolutionary tracks that graph changes in luminosity (or absolute magnitude) and surface temperature (or spectral class). Astronomers "age" stars in computers, mathematically modeling them by allowing internal thermonuclear processes to proceed to see how the stars develop. We have long known that stars stay on the main sequence for most of their lives, until the internal hydrogen fuel is exhausted, and then they brighten, expand, and cool along a particular graphical locus or track to become giants. The T Tauri stars, however, fall outside the graphical realm of the giant stars, and moreover they are clearly much too young to be giants. Instead of moving *off* the main sequence, they must be moving *onto* it. We have found not just young stars but "protostars," more accurately pre–main-sequence stars, stars in the actual process of development toward long lives on the main sequence.

Rarer than T Tauri stars and of higher luminosity are the similar Ae/Be stars, stars of spectral classes A and B that also radiate emission lines (hence the appended "e"). Their positions above the main sequence, emission lines, related reflection nebulae, and infrared emission from heated dust all indicate that they too are protostars. We thus see a continuous sequence of pre–main-sequence stars, the lower luminosity T Tauri stars logically becoming those on the lower main sequence like the Sun, the Ae/Be stars eventually producing the higher mass dwarfs.

Optical radiation cannot penetrate the thick dust of the molecular clouds, but long-wave infrared radiation passes easily through the dirty gas. With infrared detectors we can see hundreds of optically invisible stars. They must be new, the results of pre–main-sequence evolution, else their motions would have taken them out of the clouds long ago.

We would like a theory showing us how the protostars develop and mature, but a complete theory cannot be created from first principles: scientific theories are driven by experiment or observation. To see how stars are born we must identify yet earlier stages of infancy and string all the stages together. Since the black interstellar clouds are the formation sites, we look there to see what else we can find. And find we do.

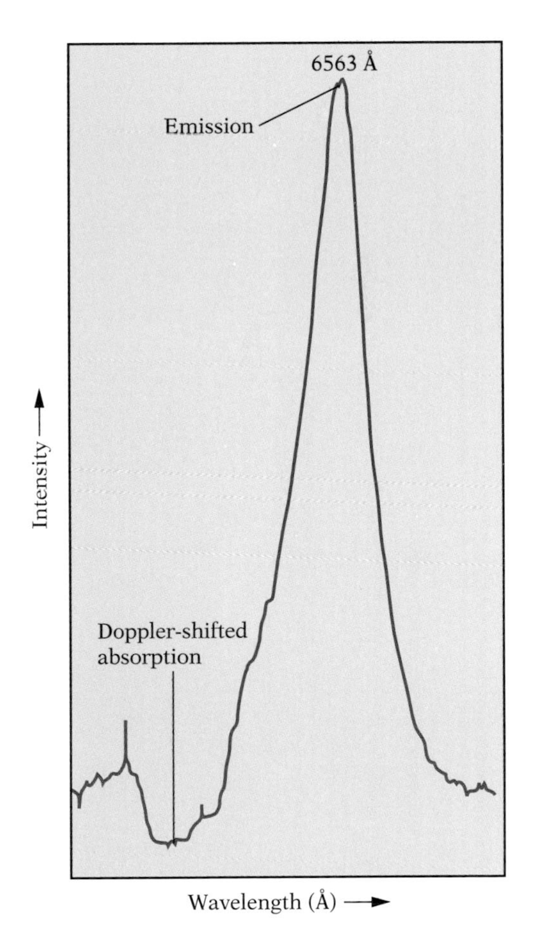

DR Tauri's Hα emission line at 6563 Å, produced by a slow wind coming from the star, is flanked to the shortward side by a Doppler-shifted H_α absorption line that is produced by gas flowing directly at the observer.

STELLAR SQUIRTS

In ordinary stars, the Hα spectrum line is seen in absorption as electrons jump upward from energy level 2 to level 3. Joy found Hα reversed in the T Tauri stars, in *emission,* the electrons jumping downward, implying a circumstellar gaseous cloud. Detailed investigations in the 1960s and 1970s revealed a more complex structure:

many of the stellar Hα lines displayed *both* characteristics, with a weak absorption feature on the blueward, or short-wave, side of the strong emission. Such lines are vividly seen in the spectrum of P Cygni, a barely naked-eye supergiant that is losing mass at a fierce rate. The outflowing gas produces strong emission lines, but the gas that comes directly at the observer in front of the star must be so dense (and hence opaque) that it creates an absorption line Doppler-shifted to the blue because of the gas's velocity of approach. Such "P Cygni lines" always mean powerful mass loss and thick stellar winds; those from T Tauri stars flow at a rate of some 10^{-7} solar mass per year, nearly a million times the flow rate of the solar wind.

The youth of the T Tauri stars shows that they must have formed recently from the interstellar medium. For them to exist at all, something is causing the dusty, low-density interstellar gas to condense into high-density stars. These new stars would thus be expected either still to be gathering mass from the interstellar medium and gaining in size, or they should at least have settled down to become more like the Sun, with perhaps a quiet, modest solar-type wind. But here we see the stars fiercely *losing* mass—a total surprise.

A subset of T Tauri stars, however, displays *inverse* P Cygni lines in which the absorption is on the long-wave or redward side of the emission; these lines show clearly that mass is falling starward and accreting as expected for a growing star. T Tauri stars in general also emit ultraviolet radiation far in excess of that allowed by a thermal gas at the observed surface temperatures; this observation is consistent with matter raining onto the stellar surfaces at high energies, matter so dense that it occasionally produces the inverse P Cygni lines. To the surprise of theoreticians, these new stars were accreting and losing mass *at the same time*. Mass loss must thus somehow be linked to the condensation process and crucial to the construction of theories of star formation. Why? Equally important, how?

The initial clues had been provided in the early 1950s by George Herbig of Lick Observatory and Guillermo Haro of the University of Mexico, who independently found faint fuzzy blobs of bright matter in Orion that had no obvious sources of illumination, no embedded nearby stars lighting them up. These Herbig–Haro, or HH, objects were complex structures exhibiting separate glowing knots that varied in brightness; in the extreme case, new knots were seen developing where none had been before. Over the following 30 years, several more HH objects were found associated with the dark clouds of Taurus, Ophiuchus, and Orion, but they remained a mystery. For a

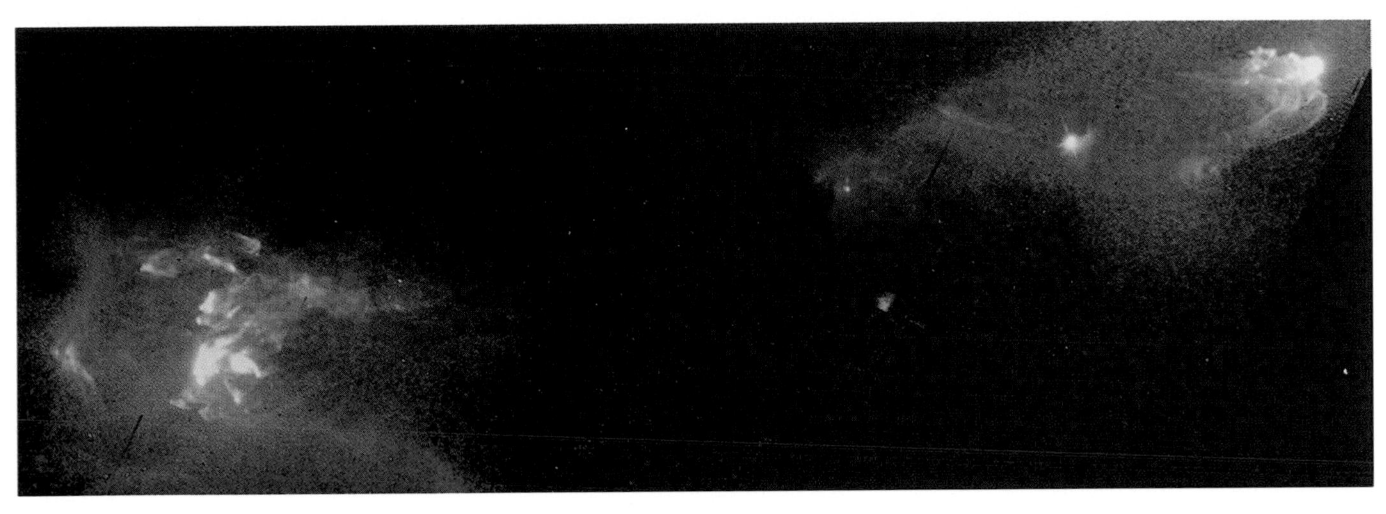

In a remarkably detailed view made with the Hubble Space Telescope, two jets stream away from a common point, at which there is a buried T Tauri star, to form the Herbig–Haro objects HH-1 (right) and HH-2 (left).

time it was thought that they were actually new stars in the process of formation.

A series of discoveries, one falling atop the other, revealed the remarkable origins of HH objects. Their nebular emission lines confirm their gaseous natures. Their spectra, however, are quite unlike those of diffuse or planetary nebulae in that they show more powerful lines of low ionization, such as [S II]. Such lines are more in keeping with supernova remnants—the debris from exploding stars—that are heated by shock waves. Shocks imply supersonic motion. But where was the energy coming from?

Measurement of the positions of two neighboring HH objects (HH-1 and HH-2) relative to distant background stars showed that they are moving in opposite directions, away from a common source: they are *related.* And in between them is a T Tauri star so buried in interstellar dust that it is optically invisible. The overall structures are huge, the HH objects a parsec or so from the star. Theoreticians quickly suggested that they are the ends of diametrically opposed gas flows squirting from the star and crashing into and shocking the local interstellar medium. Indeed, good images show that the ends of HH objects are shaped in the form of a bow shock like that seen at the bow of a moving boat. Within a year of this discovery the first jets were found, and the T Tauri winds were directly revealed as tight narrow beams of ionized matter.

How could such flows possibly be produced? The powerful infrared emission, disproportionate to the stars' temperatures, gives a clue. The amount of heated dust implied is so great that were it spherically distributed, the stars could not be seen. T Tauri stars should not be visible. That they commonly are tells us that the dust—and the gas in which it is surely embedded—is spread into a disk; if the disk is tilted, starlight can spill through the disk's poles toward Earth. The stellar wind is apparently ejected in the same direction (perpendicular to the disk) where the dusty gas is thinnest, presumably along the rotation axis of the system, to produce the opposing bipolar jets.

The stars are far away and the disks' angular diameters concomitantly small, so direct observation of them is difficult. Nevertheless, radio observations of HL Tauri's CO emission line reveal that the gas lies in a huge disk 2000 AU in radius. The velocities, found from the Doppler effect, behave according to Kepler's laws, so the gas must be in orbit around the star. Moreover, infrared interferometric observations—those making use of the interfering properties of light that allow high spatial resolution—suggest the presence of a smaller inner disk of scattering dust comparable in size to the Solar System's Kuiper belt. Even better, optical observations made with the *Hubble Space Telescope* show that the source of HH-30 in Taurus is a dark edge-on disk set within a brighter one: the star is

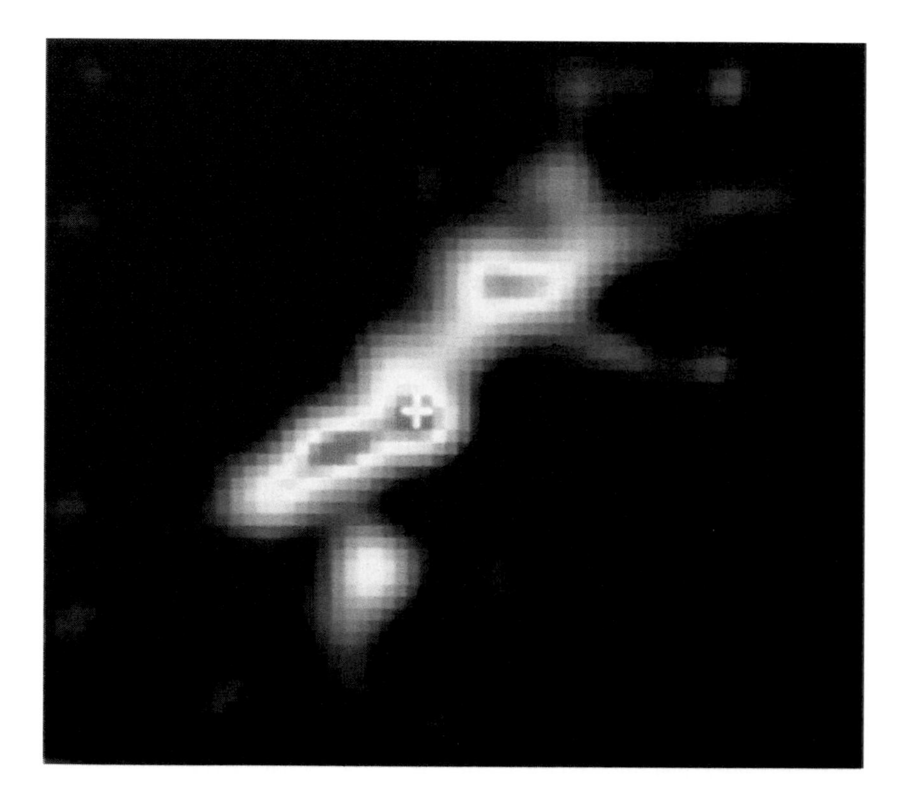

A radio view of HL Tauri taken in the radiation of the ^{13}CO molecule shows a surrounding disk of matter 2000 AU across.

completely obscured. The brightness of the edges of the disk is caused by starlight scattered outward by dust; the inner dark band is the disk itself, optically thick and only 150 AU in radius, four times the size of our planetary system. As expected, the jet that creates HH-30 emerges exactly perpendicular to the disk. These stars allow us to see what the Sun was once like, perhaps even before the planets were formed.

The disks must be the source of the gas being accreted by the T Tauri stars; they are not passive structures but active "accretion disks." Modeling of their spectra shows accretion rates comparable to the estimated mass-loss rates of about 10^{-7} solar mass per year. In some way still unclear (though there is no lack of theoretical speculation), accretion gives rise to mass loss and the bipolar jets that ultimately create the Herbig–Haro objects. The HH-30 observations show that the jets originate close to the star; were they in our own planetary system they would begin within the orbit of Jupiter, only 5 AU away from the Sun, and probably much closer. Matter shoots out like bullets about every 20 years or so, consistent with the knotty structures seen in most jets. The more gas dumped on the star, the more flees from it. The star grows because over its brief formation time input is greater than output.

The bipolar jets are thin, and though they move at a few hundred kilometers per second, they do not appear to have enough momentum to sweep up all the gas seen in the HH objects. Theoreticians hypothesized that the bright ionized flows were the centers of more massive but hard to see *neutral* bipolar flows that in a few special cases are observed directly by 21-cm emission, the ionized jets at their cores serving as tracers.

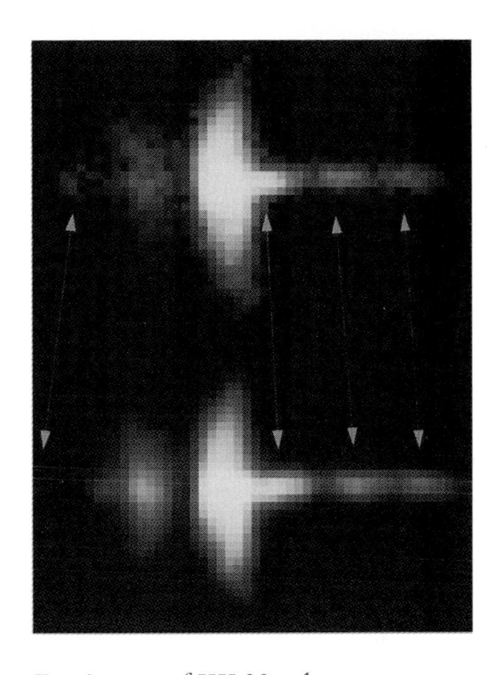

Two images of HH 30, taken a year apart, show a dark flared edge-on disk that hides what is almost certainly a T Tauri star; the disk's bright borders are visible to the Hubble Space Telescope by reflected starlight. Bipolar jets emerge from an area smaller than the size of Jupiter's orbit. Knots that reflect the accretion rate move outward at about 200 km/s, covering 50 AU—about the radius of the planetary system—in a year.

Natural Experiments

Astronomy is not an experimental science. Fortunately, we cannot adjust the interior temperature of the Sun to see what would happen. Instead, astronomers learn about the behavior of celestial bodies either by computer simulation or by looking at the full range of a class, letting nature itself do the lab work. We can learn more about the nature of the T Tauri and Herbig–Haro phenomena by observing variations on their themes.

The ultraviolet light of a T Tauri star, and much of its optical emission, is caused by gas from its accretion disk crashing onto the stellar surface. The brightness variations that first allowed the T Tauri stars to be recognized as a class must therefore be caused by equivalent variations in the stars' accretion rates, showing that the accretion process is unstable. Accretion disk masses and accretion rates must also differ considerably from one star to another to give the T Tauri stars their observed individuality.

T Tauri stars provide information about the origin of the Sun, and the Sun in turn teaches us something of T Tauri behavior. The Sun is surrounded by a vast halo, the corona, which is visible during a solar eclipse. The gas of the corona is largely confined to great loops of magnetic force that originate below the solar surface, and it is heated to some 2 million K by energy channeled through that magnetism. Like a diffuse nebula, however, the corona is too thin to be a blackbody. As a result, its luminosity is quite low (and Earthlings are not fried); the temperature is a kinetic temperature only. Coronal X-ray emission, however, which does not penetrate the Earth's atmosphere, is still quite strong and at times is mightily reinforced by dramatic solar flares that result from interacting and collapsing solar magnetic fields. Between the corona and the solar surface—the visible photosphere—lies a thin, cooler transition layer,

An optical image of the solar corona (seen during a solar eclipse), magnetically heated to 2 million K, is superimposed on an X-ray image of the solar surface. The chromosphere is a thin transition layer between the solar surface and the corona that radiates a variety of emission lines.

the chromosphere, that shines by emission lines, particularly those of hydrogen and ionized calcium, and is bright in the ultraviolet.

Magnetic fields are generated by the movement of electric charge; the most familiar example is the current in a wire. The outer layers of the Sun are ionized, thus containing free electric charges, and are in a state of convection, in which hot gases rise and cool gases fall (convection is the reason it is warmer near the ceiling of a room than at the floor). Solar convection and rotation together act as a giant natural dynamo that is ultimately responsible for the various aspects of solar magnetism. (The Earth behaves similarly, its familiar magnetic field generated by convective and rotational motions in its electrically conducting fluid iron core.) T Tauri stars display solar-type activity on a grand scale, exhibiting emission analogous to the solar chromosphere and exploding with X-ray and ultraviolet flares thousands of times stronger than their solar counterparts. By analogy with the Sun, the activity reveals powerful magnetic fields produced by a rotational dynamo.

The T Tauri stars described so far, the "classical" versions, have scaled-down analogues that have only weak emission lines and little or no infrared emission beyond that expected from the stars' blackbody temperatures. Nevertheless, these weak-line T Tauri stars produce powerful X-ray flares. They thus appear to be magnetically active stars in which accretion has come nearly to a stop, probably because the inner part of the disk has somehow been cleared; the resulting structure is that of a hollow doughnut. Could a planetary system be forming from the dusty gas?

At the other extreme is FU Orionis. Sixty years ago it was a modest 16th magnitude star—10,000 times fainter than the eye can see—that brightened by a factor of 250 over a period of less than a year. Only a few stars of this odd type are known. One, however, V1057 Cygni, was an ordinary T Tauri star before outburst (its sudden increase in brightness), and all of them are associated with reflection nebulae and Herbig–Haro objects. Presumably, all FU Orionis stars were once T Tauri stars and will become so again. From the number of known outbursts it appears that every few hundred to a thousand years a T Tauri accretion disk undergoes a major disruption, the star brightening enormously because of an increase in the accretion rate of a factor of 1000 or more. The mass-loss rate then increases at about 10 percent the accretion rate, sending large blobs of gas streaming outward; these blobs presumably are the actual origins of the bow-shocked Herbig–Haro objects. In some instances we can see bow shocks within bow shocks, memories of successive FU Orionis outbursts. Since the accretion rate is 10 times the outflow rate, the star can accrete a hundredth of a solar mass or so while in the outburst state.

Though the Herbig–Haro objects are beautifully explained by bipolar flows, the theory of how the jets and neutral flows are actually ejected is difficult, uncertain, and highly controversial. The stars are magnetically active, and magnetic fields probably pervade the disks as well. In one theory, the star has been spun up to a high rotation velocity by infalling matter and has difficulty accreting more mass. Part of the infalling mass striking the star's surface is then hurled outward equatorially above and below the accretion disk along magnetic field lines in a thick wind that can focus the observed jets to flow along the rotation axis. Disk rotation may also twist the disk's magnetic field, and the gas is then accelerated along a magnetic axis that lies along the rotation axis perpendicular to the disk. Mass loss and the jets thus probably come not so much from the star itself as from the inner accretion disk. The problem, of course, is that the action takes place too close to the star for us actually to see it.

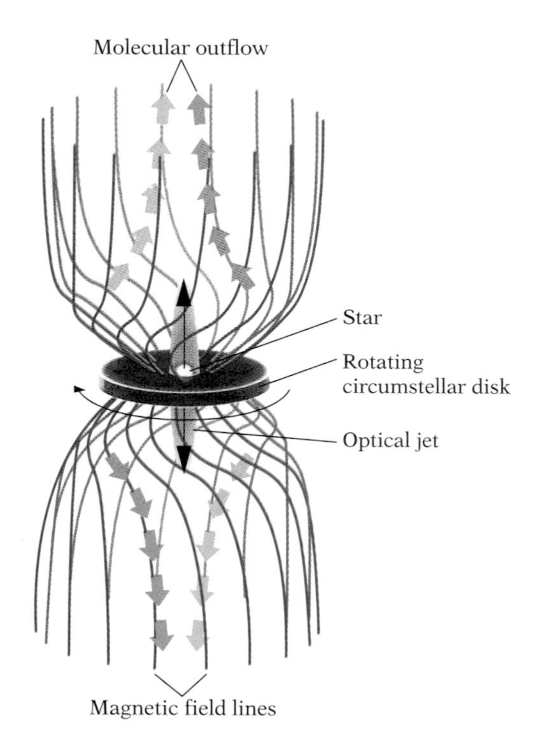

A hypothetical model of a forming star shows a rotating circumstellar disk and accompanying twisted magnetic field shooting out a fast neutral jet with an ionized central tube, the jet entraining a slow-moving molecular outflow.

Flows and Cores

About the same time the Herbig–Haro objects began to be explained by bipolar jets, radio astronomers were discovering cold molecular flows buried deeply within dusty molecular clouds. Maps of CO emission within the clouds revealed paired blobs of cold gas. One of the pair had its CO line Doppler-shifted to the blue (that is, to shorter wavelengths), the other's had its CO emission shifted to the red: the blobs were emerging from a common center. In the classic case, the flows point away from a buried infrared source called L 1551-IRS 5, the blueward flow (that pointing in our direction) containing Herbig–Haro objects. Infrared observations of IRS 5 itself reveal a jet shooting from the unseen source in the same direction as the flows and toward the HH objects. High-resolution observations of the radio continuum show the jet extending inward to only a few hundred AU from the generating source.

Different molecules provide different information. Carbon monoxide (CO) traces molecular hydrogen (H_2), which constitutes the bulk of the outflowing low-density gas. Other molecular lines, such as those of ammonia (NH_3) and carbon monosulfide (CS), trace out *high*-density molecular gas. Radio observations of CS show a disk set across IRS 5 perpendicular to the flows. IRS 5 looks very

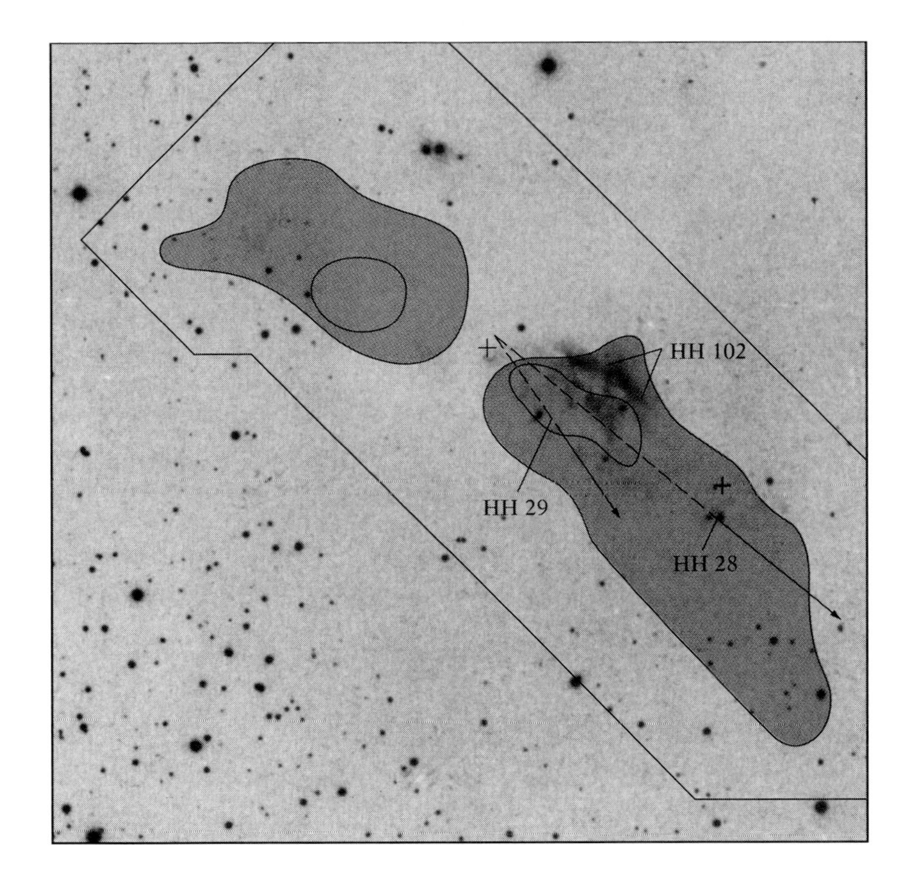

From L 1551-IRS 5 in Taurus stream two CO flows, one moving away from the Earth, Doppler-shifted to the red, the other moving forward, toward us. A jet centered on the forward flow is associated downstream with HH objects (their directions of motion are indicated here by arrows). The rearward jet is hidden by dust. Perpendicular to the flow lies a thick disk visible in CS radiation.

much like a buried T Tauri star or some kind of immediately preceding protostar.

These bipolar molecular flows are much more massive than T Tauri jets but are slower, moving at only 10 to 20 km/s. In a few instances, jets are seen at the cores of the molecular flows. The flows from protostars are apparently nested, one kind inside the other. The drivers may be the magnetically focused neutral flows seen at 21 cm, which are accented by optically bright ionized jets at their cores and which capture the surrounding molecule-rich neutral gas and drag it along. Recent observations of the HH-47 complex made with the *Hubble Space Telescope*, suggest that the shock waves within the ionized flow are the drivers, hitting and picking up the surrounding gas as they blast away from the star. Great numbers of these molecular outflows are seen embedded within molecular clouds.

There must be yet earlier stages. Observations of molecular clouds in ammonia radiation allow us to pick out "dense cores,"

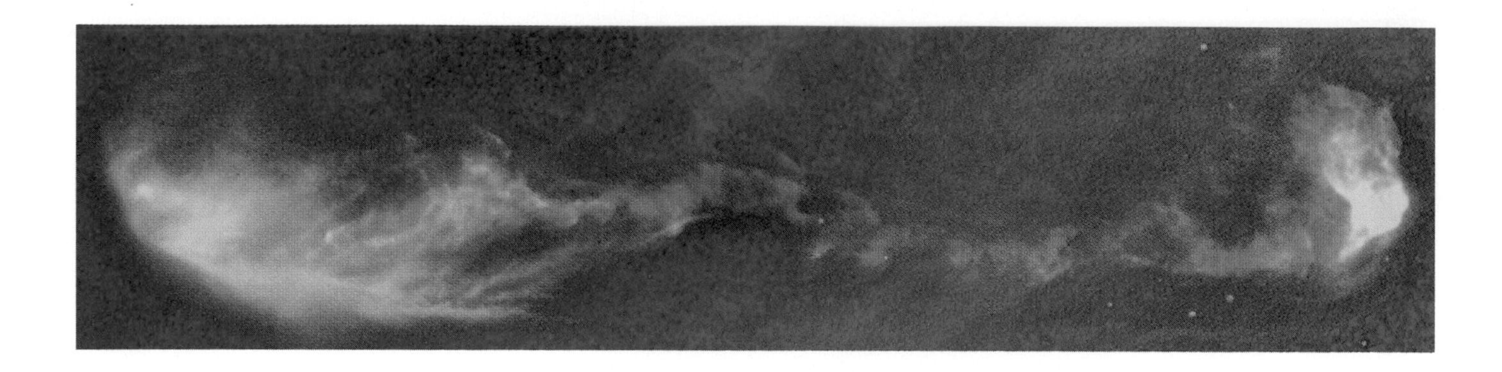

A spectacular Hubble image of the HH-47 jet shows numerous shock waves moving away from an infrared source buried in the bright nebulosity at the far left. The shocks apparently pick up surrounding neutral matter, helping to drive a molecular flow.

especially dense knots of molecular gas that contain anywhere from a few to tens of solar masses at densities of some 10,000 or so molecules cm^{-3}. But these clouds are cold and have no associated flows. Their spatial association with T Tauri stars—both kinds of objects are found in the same molecular clouds—suggests that they will develop into protostars (or even into groups of them), but have not yet even begun to heat or to form disks. Larger dense clumps could make whole clusters or associations, so the whole picture is quite complex.

The phenomena are now laid out: young stars are arrayed in a variety of forms, mass accretion giving rise to mass ejection. The next step is to string them together with a coherent theory, a theory that—albeit currently incomplete and controversial—will take us from molecular clouds to the Sun.

Recipe for a Star

> Take 10^4 solar masses of molecular gas. Sprinkle liberally with carbon and silicate dust spiced with metals. Freeze to 10 K and stir well until mixture is frothy. Hammer until lumpy. No need for oven; stars will form and bake themselves. Watch out for hot bubbles flying from pot.

The instructions just leave out a few of the details. Theoreticians, however, are rapidly filling them in, painting a picture of star formation that is logical and compelling, though complex and sometimes chaotic—with yet enough gaps to keep astronomers working for years.

We still have only a vague idea of how molecular clouds form, but the spiral structure of the Galaxy seems to be key. Spiral-arm theory is still in a state of argumentative flux; the arms seem to be giant density waves, something vaguely akin to sound waves with a 2000-pc wavelength caused by some sort of unspecified disturbance, perhaps the gravity of nearby satellite galaxies like the Magellanic clouds, or by a "bar" at the Galaxy's center. (Such bars, thick rods of stars a kiloparsec or more long that project through the Galaxy's core, are commonly seen in external galaxies.) The waves wrap up into spirals because of the Galaxy's rotation.

Like a water wave, a density wave is a flow of energy, not of matter: the passing wave compresses the interstellar medium in an irregular, chaotic fashion into distinct fragments. Cloud formation and subsequent restructuring are not very well understood, however. Molecular clouds may compress from neutral hydrogen clouds and neutral hydrogen may be cooked off the edges of the molecular clouds by hot stars; the situation is quite confused. Some clouds are not gravitationally bound and may fragment and dissipate, but others are dense enough to be tied together; still others may collide, dissipate their energy of motion through friction, and coalesce. The result is a range of cloud masses from vast complexes to small globules.

The kinds of stars born in a given cloud seem to depend largely —and not illogically—on the mass of the cloud (or cloud complex). Lower-mass molecular clouds, like the Taurus–Auriga system, spawn mostly stars of lower mass, accompanied by no or few O and B stars. Higher-mass clouds, however, have a greater likelihood of containing larger-mass clumps, and therefore can produce the Galaxy's monster stars. But the large-mass clouds are the rarer ones, and thus the O and B stars are also rare.

Stars seem to form in two distinct modes that also relate to cloud mass. In Taurus–Auriga, and to a large degree within the Ophiuchus dark cloud, they develop more or less singly out of small condensations. Thus we see scattered dense cores, some of which may associate into loose unbound clusters that will eventually be seen as T associations. In this model, little of the interstellar mass actually winds up in stars. In massive clouds, however, we also see a "packed mode" made possible by massive clumps that fragment into OB associations, the lower-mass stars following along in the clouds' outer portions. Some of these groups will be packed densely enough that their stars are gravitationally bound, resulting in open clusters like the Pleiades and Hyades. But this bimodal format may be an artifact of observation, the two forms of star formation only the extremes of a continuum.

The arms of many spiral galaxies emerge from obvious thick bars, illustrated here by NGC 1365 in the constellation Fornax.

The nearby Orion molecular clouds, which occupy the lower parts of the celestial giant (his character is appropriate to the sizes

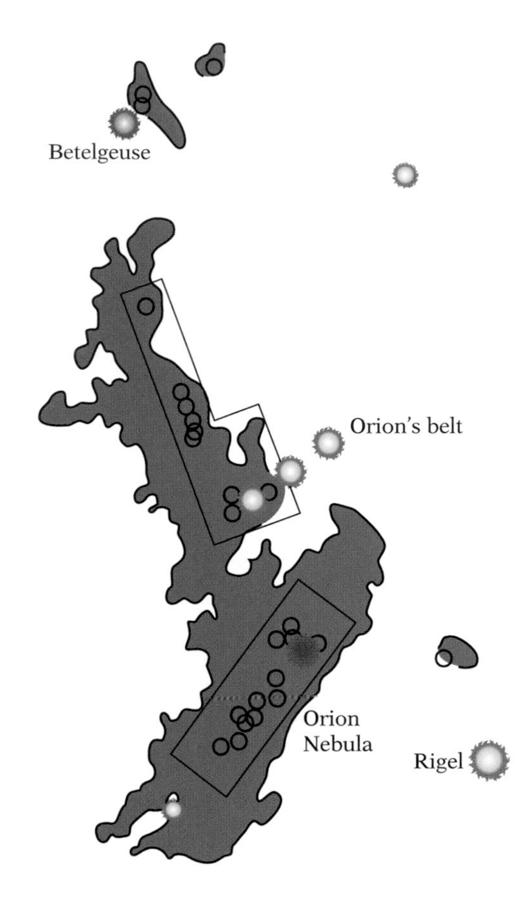

A radio view across Orion's molecular clouds reveals a number of molecular flows (circles within the boxed observed areas) that almost certainly are ejected by protostars. The constellation Orion is shown to place the radio image in context. Compare with the images on pages 77 and 119.

of his clouds), provide strong evidence for sequential star formation in which one generation of stars produces another. The O and B stars have powerful shocking winds that blow bubbles into space. The more massive stars explode as supernovae, which release even more violent shocks that are ultimately responsible for the hot component of the interstellar medium. The expanding waves compress the surrounding clouds, resulting in another round of star formation, which in turn results in another until the mass of the cloud is used up or dissipates. In Orion, a group of stars northwest of the belt seems to be the remains of the first generation. Its long-gone O stars helped create the next generation, those of the "Orion OB1 association" (OB associations are identified numerically by constellation; a second association in a constellation would be designated OB2). Orion OB1 encompasses most of the constellation, including the belt. Further compression produced by this group gave rise to the Trapezium stars that light the Orion Nebula, its youth revealed by the massive O6 main sequence star θ^1 Orionis C.

The Trapezium is centered on a fertile field of star formation. Infrared images show throngs of hundreds of faint young stars that are on or—since the region is so young—approaching the main sequence. In our part of the Galaxy there is roughly one star per 5 cubic parsecs, spaced a few parsecs apart. Within Orion's stellar hatchery stars are packed at a density of some *10,000* per cubic parsec, separated by only 0.05 pc or about 10,000 AU, a distance not much larger than the disk around HL Tauri. The packed mode appears remarkably efficient and may use up a majority of the interstellar gas and dust. Most of the stars thus formed are probably too far apart to be bound together gravitationally and will eventually scatter into the Galaxy, someday perhaps to look something like our lonely Sun.

How did our own star actually form? Its original dense core may have been created by gravitational fragmentation, in which the effects of gravity cause small density fluctuations to grow into large ones by the fragmenting effects of cloud collisions or by compression from stellar winds or supernova blasts. The ultimate driver in the creation of the Sun and other stars, however, must be gravity. Once a core is dense enough that the inward gravitational pull of its atoms exceeds the outward pressures that want to tear it apart, the core will contract, ultimately heating and producing a star inside it.

The developing star faces a major problem. All gravitationally bound interstellar clouds, and anything set within them, must have some degree of rotation, if for no other reason than as a result of collisions or close gravitational interactions. Given no outside influence, angular momentum (the sum of the products of mass, rotational velocity, and rotational radius of all the mass elements in the body) is conserved: that is, it must remain unchanged. As the average radius decreases, the average speed increases. As a result, a contracting core rotates increasingly faster. (The classic example is the figure skater spinning faster as the arms are pulled in.) The core would ultimately spin so fast it would tear itself apart, and no star could develop.

Star formation is thus impossible without the removal of angular momentum. The key to understanding star formation therefore lies in recognition of the mechanisms for that removal. The principal factor seems to be the Galaxy's magnetic field, which is most likely created by the dynamo action of the rotation of its electrically conducting gases. Since the molecular clouds are slightly ionized by cosmic rays (a vital factor in cloud chemistry), compression of the interstellar medium into denser clouds forces the field to compress as well. Though the field is weak, we can measure its strength within the clouds through the Zeeman effect, in which atomic energy levels, and thus spectrum lines, are split when radiation is emitted (or absorbed) in the presence of a magnetic field. Field strengths are weak, typically 10^{-5} to 10^{-6} that of Earth, but taken over the volume of a cloud, the energy is enormous.

The field must thread its way through the dense core. The neutral component has little use for the magnetic field, but the ions fiercely grab on to it. As gravity attempts contraction, the ions drag the magnetic field inward, the increased magnetic density producing a formidable outward pressure that counters gravity. Since the ions and neutrals are constantly colliding, that pressure is transferred to the neutrals and thus to the cloud as a whole. The cloud is then, for a while, prevented from contracting, as the outward magnetic pressure plus the natural gas pressure balance the inward tug of gravity. The magnetic force is so great that the core is forced to rotate along an axis parallel to the field direction.

But relentless gravity finally wins. The ions and neutrals are not perfectly coupled and ever so slowly pass each other, the ions sliding to the outside, the neutrals to the inside. The magnetic field thus quietly looses its grip on the contracting star, but while it holds, the field lines—still tied to the gas outside the core—grab on like ropes and slow the rotation, releasing angular momentum. A protostar is now in active formation at the center of the core.

The drawings represent different expected stages in star formation that may overlap each other. (a) A collapsing dense core is partly supported against gravity by a magnetic field. (b) As the field loses its grip, the cloud collapses from the inside out and begins to form a protostar surrounded by its accretion disk. (c) The disk and star now expel matter in a bipolar flow along the rotation axis to produce a pair of Herbig–Haro objects.

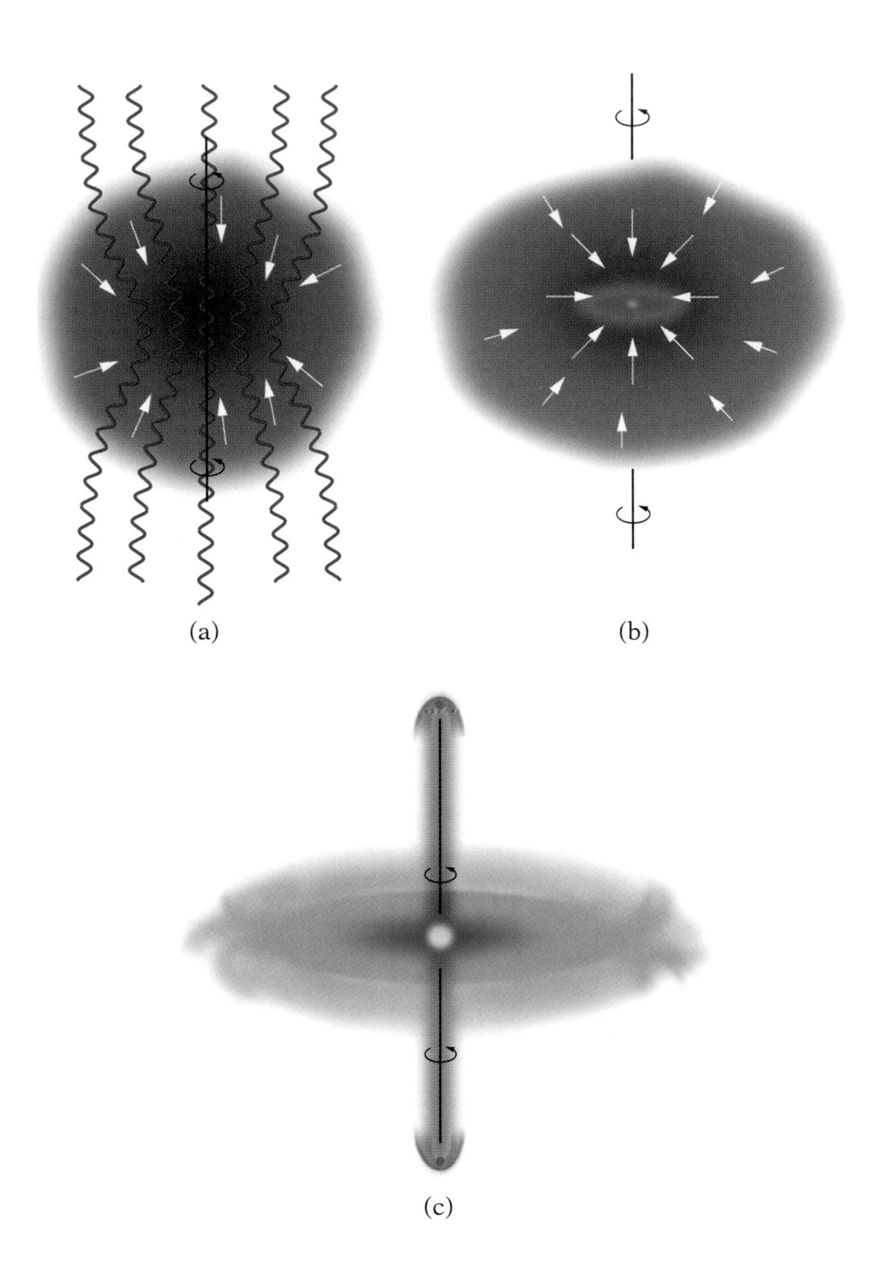

The protostar and disk are generally believed to develop by gradual accretion from the inside out, the rain of matter falling at ever higher speeds toward the inside of the core. The dusty gas near the poles, where the rotation speed is low, drops directly toward the star. But that near the equator, which is spinning more rapidly, develops into a disk, effectively a result of the conservation of angular momentum, of a contracting body spinning ever faster. An alternative view suggests that the magnetic field keeps its grip for far longer. The infalling gas slides most easily along the field lines, eventually producing a disk that rotates with the field along the core's rotation axis. Only a fraction of the core's mass is actually used to make the interior star.

The disk grows with time to radii in the tens of AU. It rotates differentially, the outer parts spinning slower than the inner. A variety of processes within the disk—convection, turbulence, instabilities caused by the disk's own gravity—dissipate energy, and mass gradually slides inward to fall from the inner disk onto the growing protostar. Rapid infall to the star heats it to temperatures high enough to make it radiate strongly in the optical. This radiation, however, serves only to heat the outer remains of the dusty core to a few hundred kelvins, allowing us from Earth to see an infrared source buried within the parent molecular cloud. The heat also destroys the molecules within the disk and alters the form of the dust grains, driving off the volatiles. The differentially rotating disk may act as a dynamo that produces a twisted magnetic field that, coupled with chaotic mass infall, launches the now-familiar jets and molecular flows that ultimately create the Herbig–Haro objects.

Most stars, however, are not single like the Sun, but double. Calculations show that during contraction the core can split in two, the additional rotational angular momentum being carried off by orbital angular momentum. The smaller resulting cores then contract to produce a binary star. Each subcore might even divide again to create a double–double, a popular format: in this case, disk creation depends on the closeness of the pair.

However a protostar is formed, its interior warms further as it continues both to accrete matter and to contract under the force of gravity. About one hydrogen atom out of a hundred thousand is in the form of deuterium (D, or ^{2}H), provided mostly by the Big Bang. Full thermonuclear fusion, the conversion of four atoms of hydrogen to one of helium that powers the Sun, requires a temperature of about 10 million K. But when the forming protostar's interior reaches only a million K, the deuterium starts to fuse, and the new energy source takes over the job of heating the interior and halting contraction (in fact producing expansion).

In a star, there must be a balance among the rate at which energy is generated, the rate at which it is transported, and the rate at which it leaves the stellar surface. Within the star's opaque interior

Computer modeling of a contracting molecular cloud shows it fragmenting into two parts, each of which will fragment into two more. The four-cloud nuclei will become a quadruple star in the form of a pair of closely spaced doubles.

gases there is a limit at which energy transportation can take place by radiation. If the internal generation rate is greater than that limit, the stellar gases begin to circulate, carrying heat along with them in a state of convection in which hot gases rise, lose their heat by radiation, fall, and pick up more. Convection constantly sweeps fresh accreted deuterium into the stellar furnace, lengthening the lifetime of this relatively stable state. This is the true moment of birth: the star is no longer a contracting blob but a viable, self-luminous, pre–main-sequence object. The stability provided by deuterium burning holds the interior temperature at about a million K, and if significantly more mass were added the star would swell to yet larger dimensions and become more luminous. These brand-new stars therefore define a locus on the HR diagram above and parallel to the main sequence called the birthline.

The newborn stars are still surrounded by active, dusty accretion disks that are still dumping matter onto the stellar surfaces. At the same time, convection and rotation in the stars can produce active stellar dynamos and flaring activity that, coupled with chaotic variations in the accretion rates, create new T Tauri stars, many still generating powerful bipolar flows and Herbig–Haro objects. At about the same time that these stars hit the birthline, the dusty shells are clearing, rendering our new T Tauri stars visible.

How the clearing actually takes place is obscure, however, as stellar winds do not seem to have enough momentum to do the job. Moreover, the bipolar flows seem to begin long before stars hit the birthline, so the correlation between the clearing of the dust and the observation of stars along it remains largely unexplained. The theories, of which there are several alternatives, all sounding reasonable

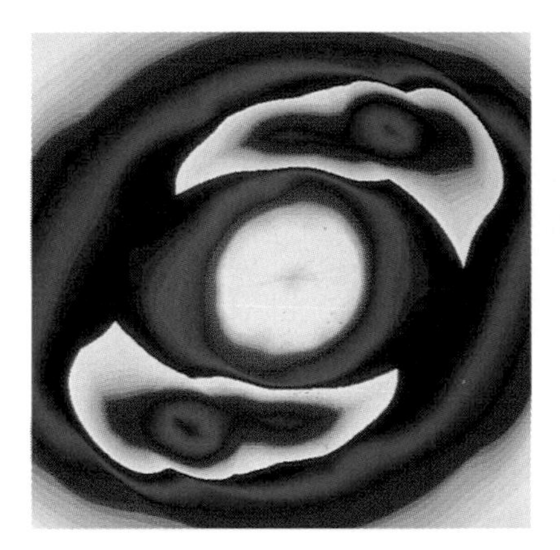
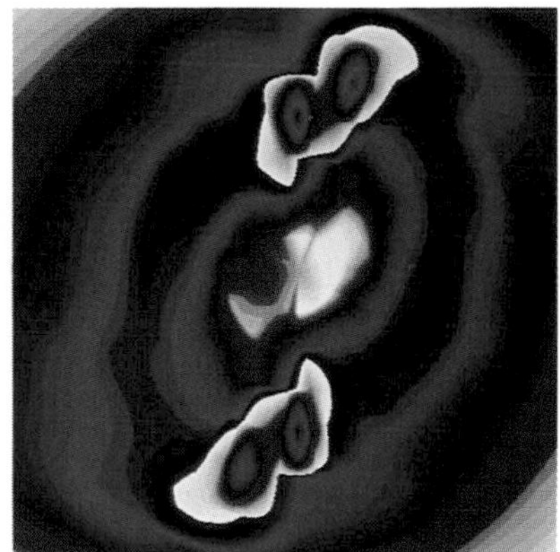

and logical, are complex, difficult, and far from complete. They will doubtlessly change faster than a T Tauri star.

The Aftermath

From the birthline, the stars descend along vertical tracks as they contract at nearly constant surface temperature. The weak-line T Tauri stars should be the evolutionary successors to the classical strong-line version as the disks clear away. But although the weak-line stars are on the average displaced toward lower luminosities, some are also found near the birthline, suggesting that chaotic processes in the accretion disks can turn off mass inflows. In any case, at some point deuterium fusion dies down, the interior of the star can transport its energy by radiation, and for stars above about half a solar mass, the evolutionary tracks turn to the left on the HR diagram as stars now heat at their surfaces at roughly constant luminosity. As the interiors heat by gravitational compression past the critical 10-million-K mark, true hydrogen fusion (starting with ^{1}H rather than deuterium) turns on and the stars quickly enter the main sequence. With a full supply of fuel, the stellar clocks are set to "zero age," the stars defining a "zero-age main sequence" and set to burn at an allotted luminosity for an allotted lifetime—both quantities depending on mass. The T Tauri stars thus become main sequence stars of classes M through F; their Ae/Be counterparts become the A and cooler B stars. For a solar-type star the whole process has taken about 10^7 years, the time decreasing as mass increases.

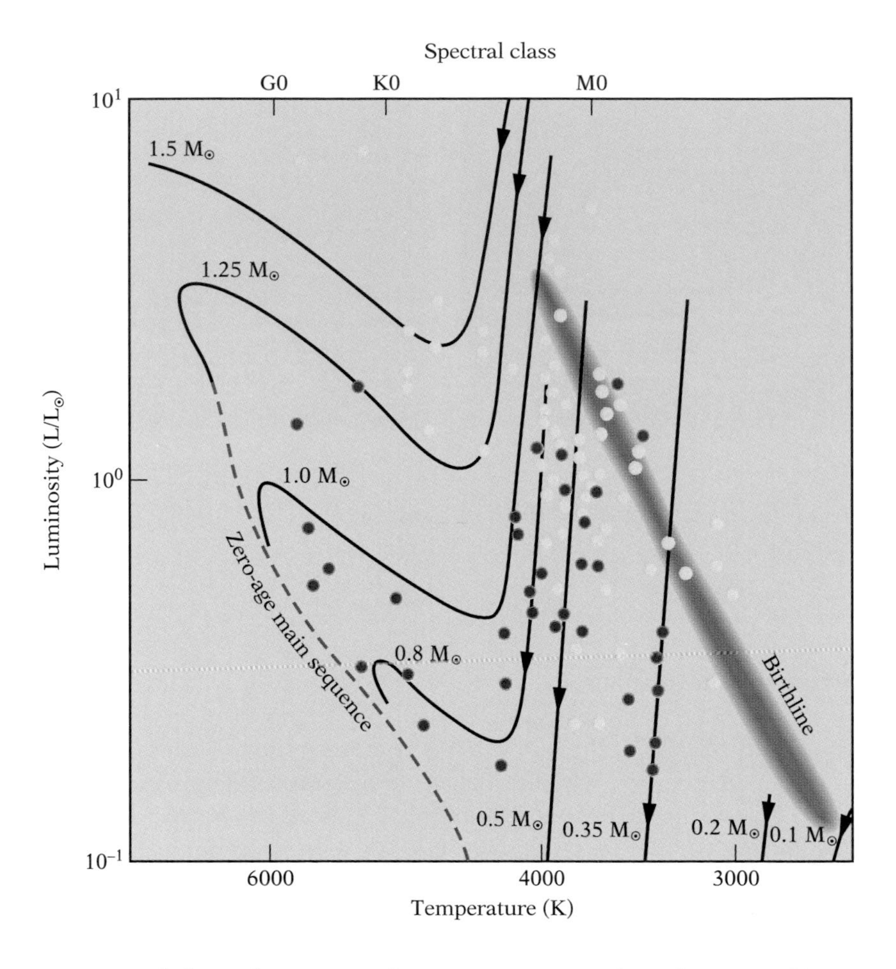

Evolutionary tracks descend from the birthline, where T Tauri stars become optically visible, toward the main sequence, where full hydrogen burning turns on. T Tauri stars—classical ones, those with all the associated characteristics, are shown in yellow, the quiet ones in red—line the tracks as they slowly become ordinary dwarfs much like the Sun.

Stars below about 0.5 solar mass contract directly vertically on the HR diagram to the zero-age main sequence. At about 0.08 solar mass the cores cannot reach the fusion point, and these stars appear at the bottom of the main sequence. Below the limit a quasi-star, a brown dwarf, can conceivably exist, burning deuterium for a while and living off gravitational energy. Though we see main sequence stars right down to the limit, only one brown dwarf discovery to date seems viable (a number subject to instant change).

The O stars, as usual, are different. Their masses are so high and the formation processes so fast that in effect they have no readily identifiable predecessors, as full thermonuclear fusion turns on before the surrounding clouds have cleared. The maximum mass is a matter of contention: prevailing opinion puts it at about 100 or 120 solar masses. Stars more massive than this, though probably physically possible, are so unlikely to be created that even in a galaxy of 200 billion stars there are none; they are just too improbable to exist. In any case, the pressure produced by the enormous radiation of such massive stars would quickly begin tearing them apart. Though we can understand the limits to stellar mass, however, we are as yet not able to understand the details of the mass distribution—why, for example, 70 percent of all stars are M dwarfs, and why, contrariwise, there seem to be hardly any brown dwarfs at all.

In spite of the remaining problems, our basic understanding of star formation seems sound. However, what about the surrounding protostellar disks? In the excitement of watching stars grow, we seem to have left them behind. What is their fate? Kant and Laplace already knew.

7

Distilling the Planets

Comet West, seen from Earth in 1976, was a dusty, long-period comet from the Oort cloud.

Although astronomers had long speculated that stars are born from the murky mists of interstellar space, no one predicted one of the great wonders associated with developing stars, the bipolar flows and jets. They are crucial to star formation, as they, or the magnetic fields with which they are intimately associated, help remove the angular momentum that can tear a developing star apart. The flows are powerfully related to the circumstellar disks, which are themselves creations of the very angular momentum that the stars need to lose, and thus the systems are self-regulating. Our planetary system seems to have formed from such a disk, and we might speculate further that, since disks are by-products of star formation, so (at least in the case of single-star formation) are planets.

We have precious few data on planets, however. Most of what we know about planetary systems comes from our own Solar System, which astronomers use as a template to understand how other systems might have been formed. The problem is that ours was created over 4.5 billion years ago. Like a nearsighted Sherlock working to solve a murder that took place in a swamp 20 years ago, we have to figure out what happened from remote examination of the final product and its disintegrating debris. With liberal application of theory we are doing just that, though at the end we may be finding that perhaps our planetary system is not quite so exemplary, as cosmic clouds continue to cough up surprises.

ANATOMY OF THE PLANETS

The eight major planets provide critical clues about their own formation and thus about the creation of the Solar System as a whole. They and the teeming debris associated with them are the bridge that links the Galaxy's protostellar disks to the solar nebula, to our own parent cloud. To understand the larger structure, we must thus first examine its details; if we can explain them, we can be confident that the grand picture of the process of our creation is accurate.

Astronomers are examining the Solar System with ever increasing intensity, using both Earth-based instrumentation and spacecraft. Our knowledge has expanded exponentially over the past few years to give us the following general picture, some of its details long known, some quite new. Neptune, 30 AU out, defines the limit to the planetary system; Pluto seems to be something else. The eight major planets fall into three broad groups, their general characteristics changing with distance from the Sun, an important clue to formation.

Earth, the natural prototype of the terrestrial planets, has a solid rocky crust, a hot plastic mantle, and a hotter still, partially molten, nickel–iron core. (Mercury, Venus, and Mars are similarly constructed, though with different proportions of the layers.) Earth is unique in having liquid water and continents that can be pushed around by mantle convection. Earth's rocky, arid Moon is at upper left.

Bunched within 1.5 AU of the Sun are the terrestrial planets —Mercury, Venus, Earth, and Mars—which are grossly similar to one another. They are small—the Earth, 13,000 km wide, is the largest among them—and have cold, solid, silicate crusts blanketed by thin atmospheres. Average planetary densities are significantly greater than those of surface rocks, implying dense metallic cores that are almost certainly iron and nickel, the most common of the metals. Volcanic action (on all but Mercury) reveals hot interiors and thick silicate mantles lying between the cores and the crusts. The terrestrial planets were constructed largely of iron, nickel, silicon, and oxygen. Some global process subsequently differentiated the elements from one another in a planetary blast furnace that sent the iron to the center, the slag to the outside. Relative to the abundances of these elements, there is almost none of the hydrogen and helium that dominate the Sun. In spite of the seemingly endless quantities of water on the Earth, the terrestrial planets are remarkably dry, consistent with their shortage of hydrogen.

Within this general framework are stunning variations. The Earth can be probed directly and in detail by measures of its gravity and by penetrating earthquake waves. Its nickel–iron core occupies half the planetary radius and a third the mass. The crust is only a few tens of kilometers thick, so almost all the remainder of the planetary mass is in the mantle. Venus seems to be similar. Mercury, the smallest major planet, has a relatively monstrous iron core topped by a thin mantle; it is not much more than an iron ball with dirt on it. Mars, the second smallest, has a minor core that occupies only 5 percent of the mass. Thus even within the terrestrial gang we find traits that are dependent on distance from the Sun: planetary radii become larger from Mercury to Earth, then decrease to Mars (which has only half the Earth's radius), and the relative dimensions of all the cores decrease outward. To these four terrestrial bodies, we might also add the Earth's Moon. Though smaller than Mercury (its diameter is about a quarter that of Earth), it is still of substantial size; it is distinguished from the others by a tiny—possibly even absent—core.

Several other features are unique to Earth. It has a strong magnetic field, the result of circulation in a liquid metallic, electrically conducting, outer core. Of the other terrestrials, only iron Mercury has a global field, and that very weak; the cores of the others are all likely solid. In part because it is the largest terrestrial, Earth contains the most heat. Convection in the solid but plastic mantle has broken the crust into several separate plates and drives them across the surface, making the continents drift relative to one another. Where plates collide we find scenes of violence: mountains, volcanoes, and earthquakes. None of the other planets exhibits such continental drift. The Earth also contains the most water, a large part of it in liquid form, which is mostly pooled into oceans that, with an average depth of only 2 km, constitute little more than a damp overlying film. Mars has some water frozen into its soil and polar caps, but Venus, Mercury, and the Moon are arid. Earth's atmosphere is uniquely rich in oxygen (one-fifth O_2, four-fifths N_2); the atmospheres of Venus and Mars are almost all carbon dioxide. (Little Mercury and the Moon have effectively no atmosphere at all.) Finally, the Earth alone has life (though Mars may once have harbored it), a feature closely coupled with its other unique characteristics.

The other terrestrial planets have their own distinguishing features. Venus's atmospheric pressure is 100 times that of Earth's, resulting in a greenhouse effect whereby CO_2 traps solar heat as a result of its infrared absorption bands; this process has raised the surface temperature to 740 K, still cold by comparison with the mantle. The planet's proximity to the Sun drove a positive feedback cycle in which high temperature evaporated water and baked carbon from the rocks, which further increased the temperature, and so on. Sulfur driven off the surface resulted in an optically impenetrable cloud deck made mostly of sulfuric acid droplets. The heat softened the rocks, allowing increased volcanism that probably chilled the core. Relative to Earth, Venus spins backwards, with a ponderous (relative to the Sun) period of 117 Earth days.

Because of its small size and weak gravity, Mars has an atmospheric pressure only 1/100 that of ours, so the planet is cold, and its water and even some of its CO_2 are frozen. Though Mars's crust is a single plate, an enormous crustal crack and huge volcanoes testify to ancient geological activity. Still, Mars is eerily Earthlike, with a 24-hour rotation period and a 24° tilt to its axis that gives it seasons quite similar to ours. Though the planet is now dry, water once flowed, empty riverbeds implying a once-thick atmosphere.

All the terrestrial planets bear the scars of meteoric strikes. Impact craters range from small holes in the ground through depressions hundreds of kilometers across, and at the upper end are giant basins thousands of kilometers wide. Earth's impact sites have been largely obliterated by erosion and geologic activity, but on Mercury and the Moon the lack of activity, as well as the lack of erosive water and air, has preserved them. Radioactive dating of the lunar surface shows that most craters and giant basins are at least 4 billion years old, created during a time of heavy bombardment that reveals an awesome violence related to the Solar System's birth.

The second group of planets, the Jovians, has two members that are nearly twins. Jupiter and Saturn are both distant, 5 and 10 AU from the Sun respectively, and huge. Jupiter's diameter is 11 times Earth's and its mass 300 times, greater than that of all the other planets rolled together and 1/1000 that of the Sun. Saturn is smaller, "only" 100 times the terrestrial mass. Both planets are made mostly of hydrogen and helium, their constitutions much more like the Sun's than Earth's. Their thick gaseous atmospheres are filled with ammonia clouds mixed with various hydrocarbons. The Jovian interiors can be deduced from the shapes of the planets, which depend upon spin rates and internal mass distributions, and by theoretical modeling. About 1000 km below Jupiter's cloudy surface, the molecular hydrogen of the atmosphere almost certainly becomes liquid, the layer extending about 10,000 km down. At that point the pressure is so enormous that H_2 is expected to take on a liquid metallic form that generates an enormously powerful magnetic field. Deep inside is a core of perhaps 15 Earth masses made of "ice," a generic term for water, methane, and ammonia whether in the solid or liquid state, and "rock," a comparable expression for silicates and iron. Saturn is similar. The huge sizes of the Jovians allow them to generate their own energy, so they radiate on their own almost as much heat as they get from the Sun as a result of gravitational contraction and, in the case of Saturn, also from the condensation of internal helium from a gas to a liquid.

Beyond Saturn lies the last group of major planets. Uranus and Neptune are distinctly different from Jupiter and Saturn. Their masses, each about 15 times that of Earth, would expand almost to the size of Saturn if they were made of light H_2 and He; however, their radii, only four times Earth's, indicate that they are composed mostly of rock and volatile ice. They are more like the cores of Jupiter and Saturn without too much extra hydrogenic baggage. Although twins, they differ from each other in two respects: Uranus

Jupiter, the prototype of the Jovian planets, displays complex ammonia clouds and the storm-like Great Red Spot, seen here toward the bottom.

is tilted 98° from the orbital perpendicular, whereas Neptune is tipped more like the Earth; and Neptune produces heat like Jupiter but Uranus does not.

The three groups are also distinguished by their satellites. Planetary satellites are either "regular" (formed with the parent planet) or "irregular" (captured by the planet). The terrestrials have no regular satellites. Mars's two moons appear to be captured asteroids, and the Earth's Moon was likely formed in a giant collision. Jupiter, however, has a wonderfully regular system that behaves like a miniature Solar System, as Io, Europa, Ganymede, and Callisto (and five others close to the planet) orbit in the planet's equatorial plane. The big four are comparable in size to the Moon and Mercury and bear the same relation between volatile content and distance from their parent body as do the planets from the Sun. The innermost of them, Io, has few volatiles (and an iron core), Europa more volatiles, and the outer two are half water ice. Beyond these are distant captured irregulars. Saturn also has a fine regular system, dominated by Titan, which, like Ganymede, is bigger than Mercury; here we find even a larger percentage of ice. Saturn also has countless satellites in the form of a set of magnificent rings that consist of billions of orbiting icy rocks that appear as continuous sheets. Uranus has five medium-sized regular satellites and some small ones, but no big ones. Neptune, on the other hand, has large Triton, but it orbits backward, and therefore must be irregular.

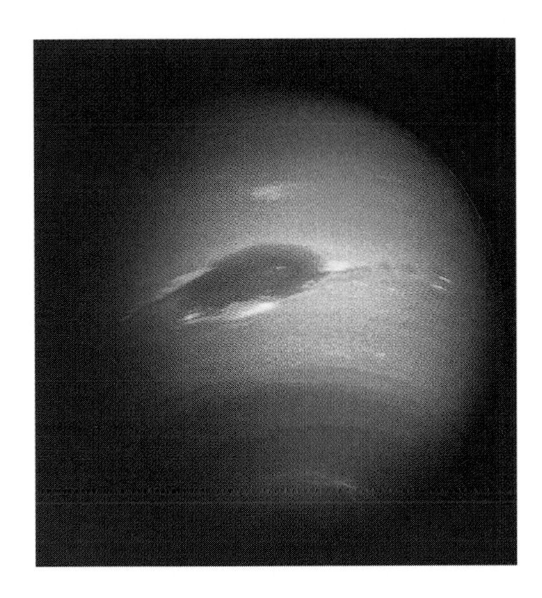

Neptune is surrounded by a thick methane haze that gives it (and its near-twin Uranus) lovely blue-green colors. The Great Dark Spot, seen here in an image taken by Voyager 2 in 1981, has since disappeared.

Pluto's orbit is distinctly odd, inclined by 17° to the Earth's orbital plane; the planet swings between 30 and 50 AU from the Sun. It is trapped into a gravitational resonance with Neptune, orbiting 1.5 times for every Neptunian revolution. Moreover, Pluto, traditionally the ninth planet, is tiny, only half the lunar diameter, and physically a near twin of Triton, Neptune's big moon. Neptune caught both bodies, which are rockier than the Jovian and Uranian satellites. They appear to represent some different kind of minor planet, and there may be more of them.

Space Junk

A variety of other, small, bodies crowd the Solar System, yielding additional clues to its formation. The asteroids are the largest, although the biggest of them, Ceres, is only 1000 km in diameter, smaller even than Pluto. Though they are concentrated between the orbits of Mars and Jupiter at an average distance of 2.3 AU from the Sun, their distribution spreads inside the Earth's orbit. Tiny ones regularly crash to Earth as meteorites (rare bigger ones occasionally

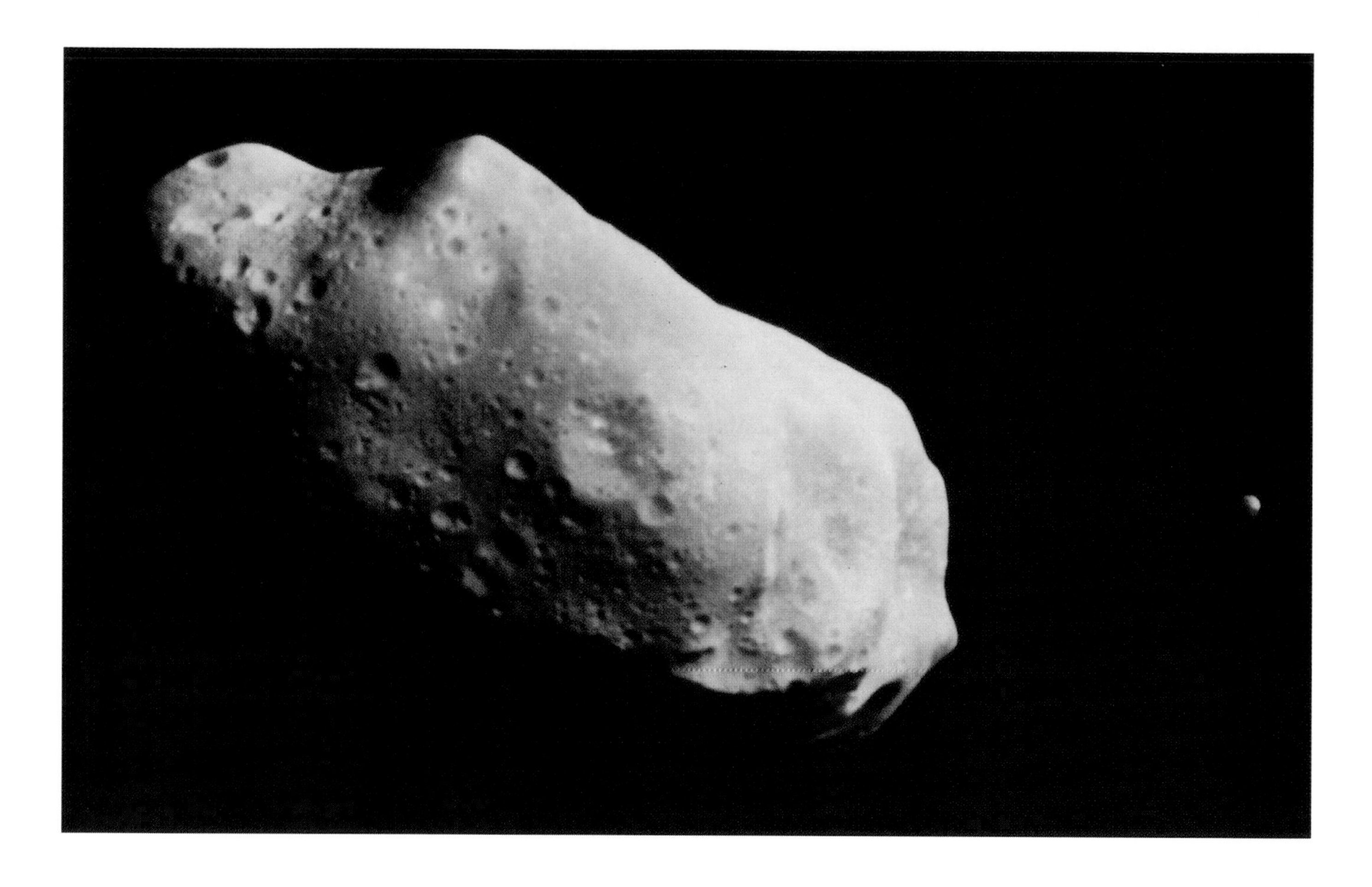

Ida, a 55-km-long rocky asteroid, orbits the Sun at a distance of 2.9 AU. Collisions have broken and cratered it. Just to the right, the Galileo spacecraft, on its way to Jupiter, also imaged a tiny satellite.

doing awesome damage), providing extraterrestrial matter that can be examined in the laboratory.

Meteorites, and thus their parent asteroids, exhibit great variety. While 80 percent of them are various kinds of stones not unlike those in a farmer's field, a small percentage are nearly pure nickel–iron, and a few, called stony irons, are a mixture. Asteroids derived from a variety of larger bodies some of which, like the Earth, had differentiated to iron cores and rocky mantles and were then broken apart through repeated collisions. The total mass of the asteroids, however, is only a fraction of Earth's, far too small to make a respectable major planet.

The stones are essentially silicates that incorporate various metals. There are many kinds, which can be arranged in a more or less continuous sequence of chemical composition in which the volatile content decreases from solar, showing a variety of sites and temperatures of formation and degrees of processing. About 90 percent of meteorites have not been much altered or processed; these, known as chondrites, contain millimeter-sized spherical stony inclusions called chondrules that look as if they had been flash melted and then frozen before incorporation into the surrounding matrix. A subset of the chondritic meteorites contain water and have high carbon contents; these carbonaceous chondrites appear to be the most primitive bodies of the Solar System. Their inventory of volatiles shows that they must have been formed from the solar nebula at temperatures of 1200 to 1450 K. The chondrules, however, show the effect of flash heating to higher temperatures that took place on the order of seconds, turning them into droplets that froze over the course of only hours; they were then incorporated into the silicate matrices. Some chondrites also contain centimeter-sized calcium-aluminum–rich inclusions—CAIs—that are very rich in refractory elements. These have been heated and cooled over a period of perhaps days, even months. The CAIs are surrounded by micron-thick rims that, like the chondrules, were flash heated to 1600 K or so. Chondrules and CAIs, with ages of up to 4.57 billion years, are the oldest things known. Meteorites thus show that the early developing Solar System was a place of steady formation—one kind of body being incorporated into another—punctuated by a variety of sudden events. The pattern is no surprise considering the chaotic, highly energetic jets and flows that emerge from the disks of forming protostars.

Meteorites contain a surprising variety of odd trace constituents that further indicate where they came from. We see what appear to be raw tiny interstellar grains. Consistently, unprocessed meteoric matter contains tiny diamonds, trillions per gram. Thirty percent of interstellar carbon grains may be in the form of these tiny crystals, more firmly linking the Solar System to its ultimate origins in dark, dusty molecular clouds. Meteorites also contain daughter products of radioactive isotopes with such short half-lives that they no longer exist within the Solar System. Small amounts of the magnesium isotope ^{26}Mg reveal that the grains used to contain aluminum in the highly radioactive form ^{26}Al, and xenon isotopes show the one-time presence of plutonium. Some of these long-gone isotopes may have been created within the forming Solar System by intense particle radiation from a new, very active Sun, but others, like plutonium, could have been created only in the furnaces of exploding stars, taking us back to the environment within which our Sun was created. Moreover, primitive meteorites contain more than 40 different kinds of amino acids—biomolecules. Complex organic molecules, including one amino acid, glycine, also float within the star-birthing giant molecular clouds. The original, interstellar, biomolecules were almost

A stony chondrite is cut to show its small chondrules. Meteorites contain a complex mixture of substances, including diamonds that may have come directly from interstellar space.

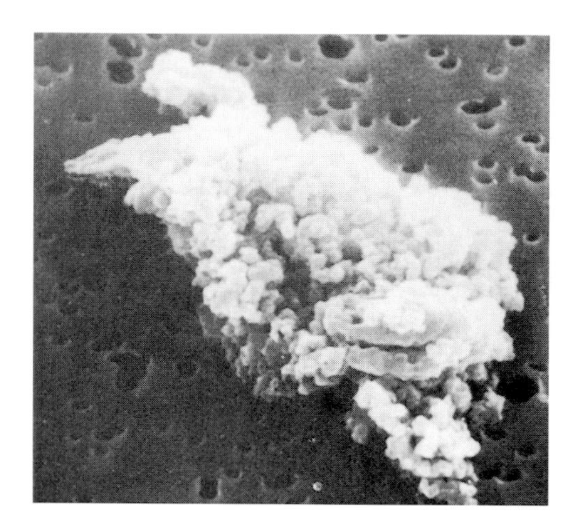

This "Brownlee particle," collected from the Earth's atmosphere by aircraft, was probably a tiny piece of some anonymous comet's nucleus.

certainly destroyed in the heat of star formation, but the interstellar version provides a potent guide for showing how our meteoric biomolecules could have been created.

The meteorites, and thus the asteroids, relate strongly to the terrestrial planets, tying these to the interstellar medium as well. They are made of the same stuff as the terrestrials, and some were processed and differentiated in much the same way at similar temperatures. They tell us a great deal of how the terrestrial planets were put together. The comets relate more strongly to the giant outer planets. The ices of which comets are largely made are complex and include water (which is dominant), carbon dioxide, formaldehyde, methanol, acetylene, methane, ammonia, and many other chemicals. Infrared spectroscopy reveals complex grains made of carbon, hydrogen, oxygen, and nitrogen. Volatile ratios are similar to those found in molecular clouds. Comets must have been formed at low temperatures, between 20 and 60 K, and have not been processed very much. Such temperatures should have prevailed only in the outer fringes of the solar nebula, beyond Jupiter, perhaps only beyond Saturn well into the neighborhood of the Uranian planets. But comets are no longer much found within the planetary system. Creatures of the distant reaches of the Sun's family, they now outline the disk-shaped Kuiper belt and fill the giant Oort cloud. These are the clues, from which we try to understand the present state of the Solar System and the story of our birth.

From Dust to Planets

One path leads from the dusty gas of interstellar space to the planets of today. We have found pieces of it, small friendly trails in the wilderness, but have yet to connect them. Along the way there are potholes and, worse, forks in the road. We travel here the most likely route—but it is necessary to keep in mind that some of the turnings may be wrong and that parts of the road will need resurfacing, if not complete reconstruction, as new discoveries are matched to theory.

Return now to the early Sun, a T Tauri star encircled by a rotating gaseous disk, the solar nebula, perhaps hundreds of AU in radius. The nebula is filled with metal-containing silicate and carbon dust from the parent molecular cloud. If all of these were broken back to atomic form, the disk would have the same chemical composition as the Sun. Heat generated by the formation of the protostar probably destroyed much of the molecular gas, but in the now-chilled disk the molecules can be re-created by much the same

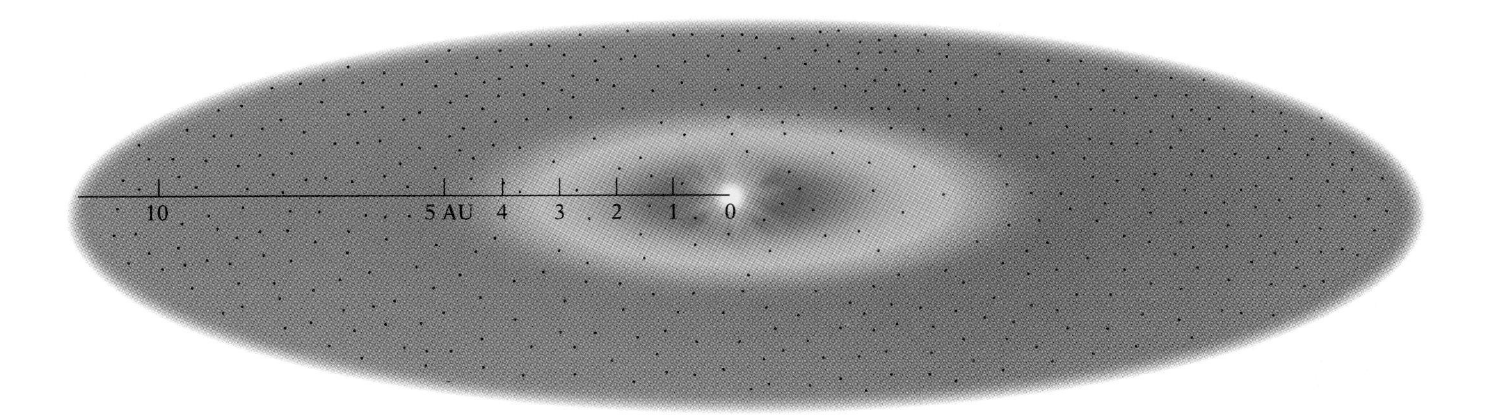

The temperature and chemistry of the disk around the new Sun, which circulated counterclockwise, changed radially outward. Near the Sun, within a distance of 1 AU, the temperature was 1200 K or more, preventing any volatile compound or element from condensing into or onto the tiny embedded dust particles. Proceeding outward, temperature decreased and more and more volatile substances could be incorporated. When the temperature dropped below about 150 K, beyond 5 AU, water and other volatiles could condense as ices, the water content of the grains increasing outward.

processes as they were in the interstellar medium, yielding a dusty disk filled with an intricate chemistry. The gas and dust are in some sort of chemical equilibrium that depends on the local temperature of the mixture and on the pressures of reactive oxygen and hydrogen. New grains can condense out of the gas, and these and the ambient ancient interstellar grains grow by the accretion of free atoms and molecules. Far from the Sun, tens of AU away where the temperature is low, below 150 K or so, volatile elements and molecules—carbon, water, alcohols, and many more—can be incorporated into the grains. But within about 5 AU, the temperature of the heated disk rapidly climbs, reaching as high as 1400 K or so near the Sun, and volatiles are increasingly destroyed or kept in the gaseous state, the inner grains thus becoming richer and richer in refractory elements and compounds.

To a simple approximation, the grains are in Keplerian orbit around the Sun but also feel powerful dynamical forces from gas drag and the absorption of solar radiation, both of which act to bring the particles inward. Grains at similar distances from the Sun move relatively slowly past one another, so many of them that they come into contact and sometimes stick together because of their atomic electrical forces. Larger particles thus grow from smaller ones, from micron to millimeter size; they are watery (icy) far from the Sun, beyond Jupiter, waterless (iceless) near it. The disk temperature of the inner Solar System, inside the orbit of Jupiter, is just right to produce the raw materials of the chondritic meteorites.

Meteorite inclusions suggest that the disk near the Sun, again that portion of it within about 5 AU, may have been a scene of terrible violence. Gravity should pull larger particles toward the center of the disk, and particle separation perhaps leads to charge separation. A dramatic (though certainly controversial) result may have been huge electrical discharges—bolts of nebular lightning—that may have been a fraction of an AU long. Growing particles caught in the blast would have melted, their volatiles driven away, and their small sizes would have allowed the droplets to freeze very quickly. An alternative theory suggests that the sudden heat source may have been intense magnetic outbursts on the new Sun. As the sizes of the particles increase by further accumulation of dust, the solidified melts—however they were created—would be buried, to appear as the chondrules in today's meteorites. The origins of the CAIs are unknown, but they appear to have required local temperature increases to 1800 K and thus to have been created relatively close to the Sun. The flash-heated rims on the CAIs also indicate that violent processes were at work. Farther from the Sun, beyond about 5 AU, where the temperature is low, there is insufficient energy to produce chondrules, so the nascent icy dust that will produce the comets is free of inclusions.

The particles continue to grow by constant collision and accumulation, and their number decreases. But gas drag and turbulence within the disk keep the particles, now larger, crossing one another's orbits and maintain the rates of growth and collision. From these simple beginnings, the particles grow in diameter to centimeters, then to meters, eventually forming trillions of real planetesimals a kilometer or more across. Just how the coagulation proceeds, especially in the centimeter-to-meter range, is still problematic, as collisional energies would have been great enough to shatter the particles back to dust grains. But coagulate they must have done for the larger bodies to be here today.

At this point, gravity becomes important. Larger bodies attract smaller ones, and (as in so many things) the bigger win. Larger particles drag the smaller ones into themselves in ever more furious collisions. Within a few hundred thousand years, the countless dust particles of the original cloud have grown through the planetesimal stage, in which they were bodies a few or a few tens of kilometers across, to a new category, protoplanets—hundreds of them, maybe a thousand kilometers wide, about the size of Ceres, the largest asteroid. These bodies continue to collide, the bigger ones again winning in a pattern of runaway growth as their gravity perturbs nearby protoplanets into crossing orbits and causes the smaller ones to be gobbled up.

In the inner Solar System, the end result of the particle accumulation and steady collision, as seen in numerous computer simulations, is a few—perhaps three to six—rapidly developing terrestrial planets devoid of volatiles within a few AU of the new Sun. Mutual

Huge bodies crashing into the early Earth and the other nascent terrestrial planets raised their temperatures high enough to melt and differentiate them.

perturbations among remaining protoplanets and residual gas drag continually bring them into the gravitational "feeding zones" of the larger planets, which absorb them violently, at fierce speeds at least as great as the new planets' escape velocities. Orbit-crossing is so severe that the inner Solar System becomes fairly well homogenized, the effect of the original temperature gradient—the variation of temperature with distance—of the solar protoplanetary nebula now long gone. The impacting bodies bury themselves deeply as they vaporize, depositing vast amounts of heat far below the planetary surfaces and causing the planets to begin to melt. The heavier iron and nickel fall to the interior and the lighter silicates float to the outside as the planets differentiate. How the primitive asteroids differentiated remains something of a puzzle: they seem too small to have succumbed to collisional heating. Heat provided by radioactive ^{26}Al and powerful magnetic solar activity may have played important roles.

Truly massive collisions take place among the largest protoplanets on crossing orbits. Mercury's small mantle cannot be explained only by its proximity to the Sun; a collision that ripped away much of the rocky covering seems mandatory. Our own early Earth seems to have been struck by a Mars-sized body, the impactor's iron core merging with ours and the collision sending into orbit a giant gaseous silicate plume that later condensed to become our Moon. The theory explains the lack of lunar volatiles relative to Earth, the serious depletion of iron, and the lunar orbit, which lies in the plane of the Solar System, not in that of the Earth's rotation. Such collisions also explain the terrestrial planets' rotation characteristics, including Venus's slow backward rotation.

The Jovian planets began to grow in the same way. In the inner Solar System, the Sun had blown away the gas. But from about 5 AU and beyond, there was plenty of it, and the icy cores of what were to be Jupiter and Saturn began to collect the gas gravitationally, becoming surrounded by vast bound clouds of hydrogen and helium. At this distance, solar radiation could not heat the gas sufficiently to drive it away. The two big planets grew until they swept up all the available gas in the disk, the solar wind and solar radiation driving out the rest. Saturn's lower mass probably reflects a lower density within the forming medium, a simple reduction of raw material. In these distant planetary reaches, however, the view is less clear.

Jupiter grew at the same rate, if not faster, than a new terrestrial planet trying to form outside the Martian orbit. Its immense gravity fired planetesimals into the region and so stirred it up that no accumulation was possible, many bodies being lost into perturbed orbits. The result was a few remaining planetesimals that collided and ground themselves down to become the asteroids and meteorites which together constitute a fragmented failed terrestrial planet. Mars, farther away from Jupiter, developed into a real planet but remained small because of a similar lack of raw material.

Beyond Saturn, the cores of Neptune and Uranus grew more slowly because the density of icy planetesimals was lower, and by the time they could gravitationally accumulate much gas the Sun had somehow dissipated it. As a result, the two Uranian planets did not become as large as Jupiter and Saturn. A violent collision of Uranus with some other proto-Uranian knocked it on its side. Most of the dust should have been gone from the disk within 10 million years or so, the planets achieving their final masses within about 100 million years, consistent with the difference in age between the oldest rocks of the lunar surface and that of meteoric inclusions and supported by the rate at which dense disks around young stars disappear.

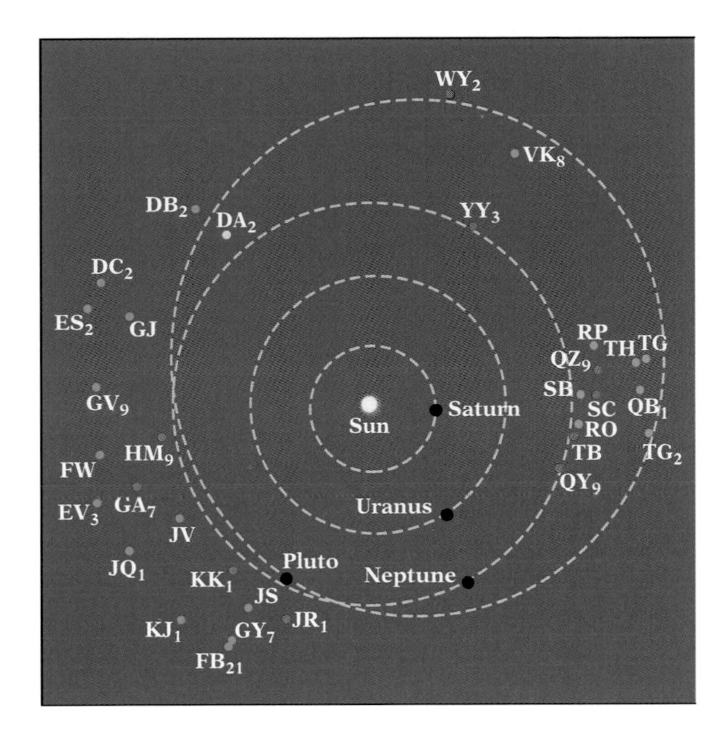

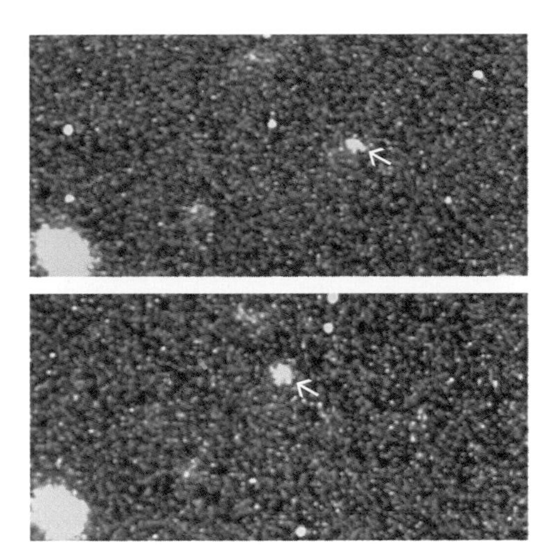

The object 1992 QB1 (above), about 200 km across and the first Kuiper belt object to be found, is moving near the orbit of Pluto. The positions of many such bodies are shown in the diagram at left; note the similarity of their distribution with Pluto's orbit.

The satellites of the giant planets perhaps formed in something of the same way, from circumplanetary disks produced perhaps by accretion or spinout of gas thrown off from the parent planet by rotation. These disks were almost certainly much denser and hotter than the solar nebula, and these conditions altered the chemistry of the gas and allowed the conversion of molecular nitrogen and carbon monoxide to methane and ammonia, making the satellites rich in these compounds as well as in the ubiquitous water ice.

Beyond Neptune, the density of icy planetesimals was so low that no planetary cores could accumulate. We are thus left with a thinly populated Kuiper belt that may extend outward for over a hundred AU, perhaps to the extent of the disks surrounding current protostars. A few bodies in the Kuiper belt could have grown to protoplanet size, the effects of mutual gravitation sending a few toward the Sun, one to be captured by Neptune as Triton, another caught into orbit as Pluto. Because short-period comets are quickly evaporated away by the Sun, there must be more bodies still out there to maintain a fresh supply, even after 4.5 billion years. Modern telescopes with sensitive electronic detectors are finding these Kuiper belt objects: so far about three dozen in the inner belt with diameters of 100 km or so have been specifically identified and their orbits calculated, and the number will increase very rapidly.

Vast numbers of scattered planetesimals of various sizes were left over, rocky ones in the inner Solar System, icy ones in the outer. Their orbits stirred by the gravity of the giant planets, asteroids and comets both were sent into the inner Solar System, where for half a billion years they, and local leftover debris, collided with the cooling solidifying crusts of the terrestrial planets. The comets brought volatiles, including water that fills the oceans of today's Earth, water that underlies the Martian hills. The Moon lost its water as a result of its collisional formation, small size and weak gravity, and lack of atmosphere, but the impact scars of the last part of the bombardment are still to be seen, as they are on Mercury and the satellites of the outer planets. When the intruders were largely swept up about 4 billion years ago, the rapid impacting stopped. A very few large bodies still hit the Moon and left obvious basins, later filled by lava to create the maria, the dark features that are visible to the naked eye. In the last 3 billion years the impacting rate has been low but steady: bodies of at least 10 g strike our planet nearly a thousand times a day.

Huge numbers of planetesimals were accelerated to the Solar System's escape velocity by Jupiter and Saturn and hurled outward into the void of interstellar space. Uranus and Neptune, however, with their lower gravities, launched the icy leftovers in their region of space—the comet nuclei—into long, but bound, elliptical orbits. Tides raised by the Galaxy then partially circularized the orbits to radii of thousands of AU, populating an inner Oort comet cloud. The masses of Uranus and Neptune, and the time it took them to form, require a total planetesimal mass 10 times the planetary mass, leading to an amazing estimate of 10^{13} comets in the inner part of the Oort cloud. Further gravitational pulls of passing stars and of the Galaxy at large gradually moved some of these objects outward into an outer Oort cloud tens of thousands of AU across that is thought to contain 10^{12} comets. There the disturbances raised by stars and giant molecular clouds can alter a few orbits into long ellipses again and rain a few long-period comets into the inner Solar System, providing an echo of the past and clues to present understanding.

Is this in fact the right path? Other routes beckon. Some astronomers think the Jovian planets were made whole from the condensing dense core, much as binary stars are made. Maybe Jupiter is a genuine brown dwarf rather than a package delivered "with some assembly required." Perhaps too the Oort cloud comets were created in place early in our history rather than launched outward by the distant planets. Little of the pathway suggested here is not a matter of scientific contention. Some theories will be confirmed by further sophisticated observation, and others will be destroyed.

HOME AT LAST

The planets, including our Earth, are thus distillates of the interstellar medium in which, to widely varying degrees, the volatiles were driven out. Small our planet may be, but it is also a highly unusual treasury of rare heavy elements. Our Earth indeed has volatile compounds, but they were mostly brought to us from the outer Solar System aboard crashing comets. We can now also see why planetary natures are so dependent on distance from the Sun: they are results of the distribution of mass in the solar nebula, of the effect of gravitational perturbations by Jupiter, or of the effect of solar heating on the planets and on the original solar nebula.

The existence of our planet is firmly linked to interstellar processes. Interstellar studies show us that complex molecules can be made in a cold, dusty gas. Amino acids are found in meteorites, apparently placed there by a similar chemistry during planet formation. Could our own origins therefore be interplanetary—or even interstellar? Could interplanetary debris have brought to Earth the seeds of life? Or did life begin here independently of impacting bodies? No one knows, and we can only speculate. Ancient fossils

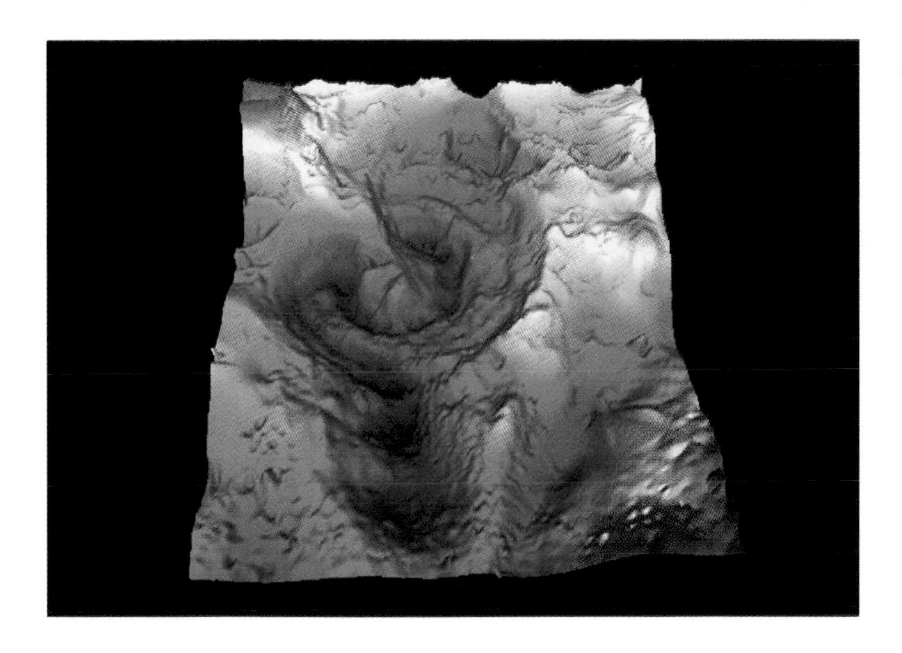

Chicxulub Crater, imaged by gravity measurements, underlies part of the Yucatán Peninsula. Two hundred or more kilometers across, it is believed to be the remnant scar of a collision with a giant meteorite that wiped out the dinosaurs 65 million years ago.

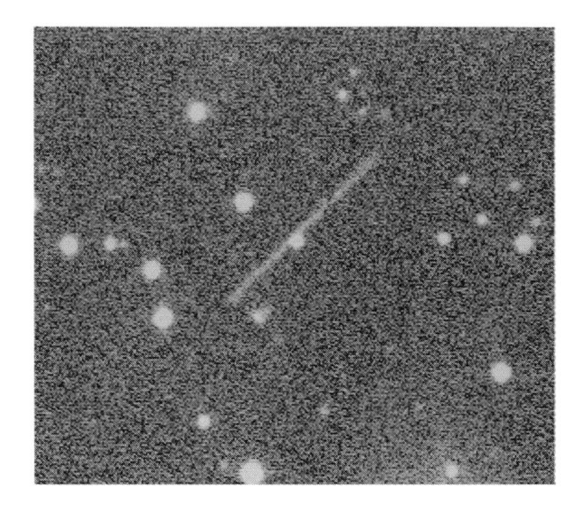

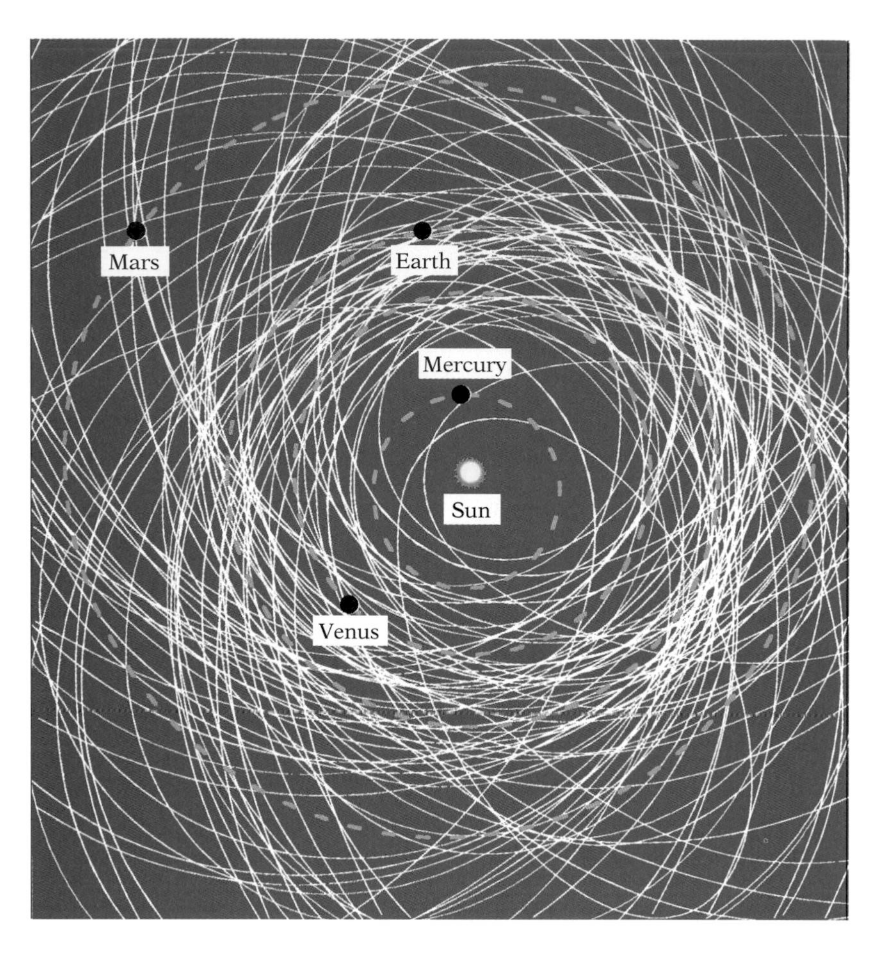

A map of the inner Solar System (at right) shows a web of the larger known Earth-crossing orbits of asteroids and dead comets. In a time exposure taken in 1995 (above), a smaller such body about 300 m across is caught moving between the Earth and Moon, only 100,000 km away.

3 billion years old show that life could hardly wait to get started on this planet. Yet why did Mars, which once had running water and a thick atmosphere—conditions apparently necessary and conducive to life—seem to have developed no forms of life? Or did it?

But whether life originated as a result of impacting debris or not, it has certainly been affected by it. Pieces of comets and asteroids continually strike us in the form of meteors and meteorites; you are breathing comet dust as you read this. Most of these particles are small, but a few are not, and occasionally they produce a big strike like the one that made Meteor Crater, a kilometer-wide hole blasted out of the northern Arizona plateau by an iron meteorite about 50,000 or more years ago. In 1908, a stony chondrite about the size of a house blew up in the air above the frozen forests of Siberia, blasting down trees for 30 km around the impact site, knocking down farmers 50 km away, and sending an atmospheric pressure wave twice around the world.

We may owe our lives to such an event. The layers in the Earth that are dated at about the time of the extinction of the dinosaurs 65 million years ago are uncommonly rich in iridium, an element considerably more abundant in meteorites than in the Earth. The implication is that the dinosaurs were wiped out by the collision of an asteroid 10 km in diameter. The impact would have raised huge amounts of dust and might have sent a shock wave through the planet that could have triggered massive volcanic flows that added to the damage. Some 80 percent of the species alive at the time were extinguished, allowing mammals—and eventually us—to take over.

Such an impact can, and will, happen again. Huge numbers of asteroids and defunct comets cross the Earth's orbit, and every year astronomers find one or more of them tens of meters across passing by us at distances comparable to that of the Moon. The Earth's cross section is about 1/1000 the cross section of the lunar orbit, so at minimum one such body big enough to do extraordinary damage should have its orbit perturbed sufficiently to take direct aim on us at least once every thousand years. There is a movement now afoot to track and alter the orbits of those homing in on us before they strike. We seem ready and able to defend ourselves from the very stuff of the interstellar medium that gave us birth in the first place.

Of Other Planets . . .

We know now that Kant and Laplace were right, that our planetary system was created from a spinning dusty disk, from the solar nebula. We see such disks around developing protostars and T Tauri stars. Will planets develop there too? Do they orbit other mature stars? What evidence can we gather that suggests other planetary systems, other distillates of interstellar space?

Finding remnant disks would bridge the 5-billion-year gap between the primitive protoplanetary disks and ours. Surrounding

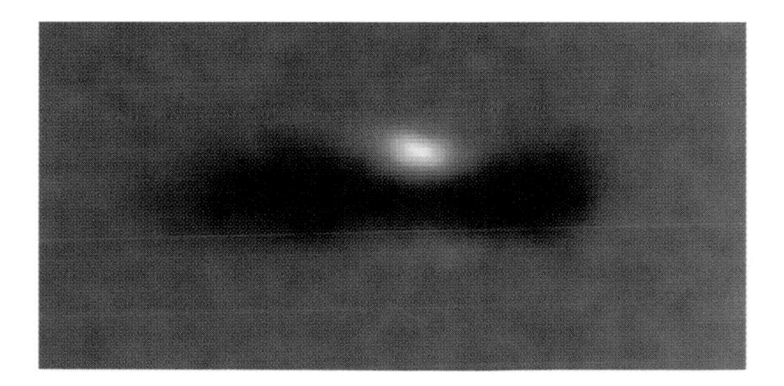

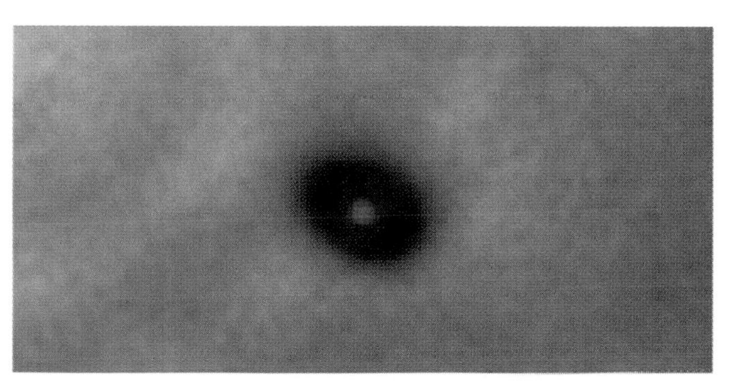

The Hubble Space Telescope imaged a number of dusty disks surrounding young stars in the Orion Nebula. The one at left is presented edge on, a bit of embedded starlight reflecting from its center. Could planets be forming even as we watch?

several of the stars buried within the Orion Nebula are disks a few hundred AU across, some of which are illuminated by the radiation from the Trapezium. Spectroscopy of a few of the stars shows that they are still contracting onto the main sequence. Surveys suggest that about half the stars within the region have such disks, which appear to have descended from those that lie around T Tauri stars.

In 1981, *IRAS* observations found that the bright, mature A-type main sequence star Vega (which passes almost overhead in northern summers) is surrounded by extended infrared emission, as were Fomalhaut and Beta Pictoris, both southern-hemisphere A dwarfs. The emission could be interpreted only as arising from surrounding clouds of warm dust particles heated by starlight. Ground-based observations of β Pictoris quickly revealed that the emission comes from an edge-on disk that reaches some 400 AU from the star. The infrared emission at different wavelengths—the continuous spectrum—can be modeled in terms of grain size, density, and temperature. The grains are larger than those in interstellar space, and a lack of high-temperature emission indicates a "hole" in the disk near the star that is actually detectable in processed images. Spectroscopy shows that the grains are made of a combination of silicates and carbon. The system looks eerily like our own. Could the disk be the remnant of planet formation, could it be the star's Kuiper belt? Is the inner portion relatively empty because planets have swept up the dust? A slight warping of the disk in fact suggests the gravitational effect of an orbiting body akin to Jupiter. Further evidence is seen in transitory absorption lines in the stellar spectrum that are interpreted as arising from comets that vaporize as they fall into the star; such comets are in fact seen to fall into the Sun.

The next step is to find the planets themselves—if they exist. Our planetary system provides a hint of how common they might be. Each of the giant planets of our Solar System has its own planetary system, its family of regular satellites. Neptune's is skimpy, but the missing satellites may have been eaten or sent off by Triton dur-

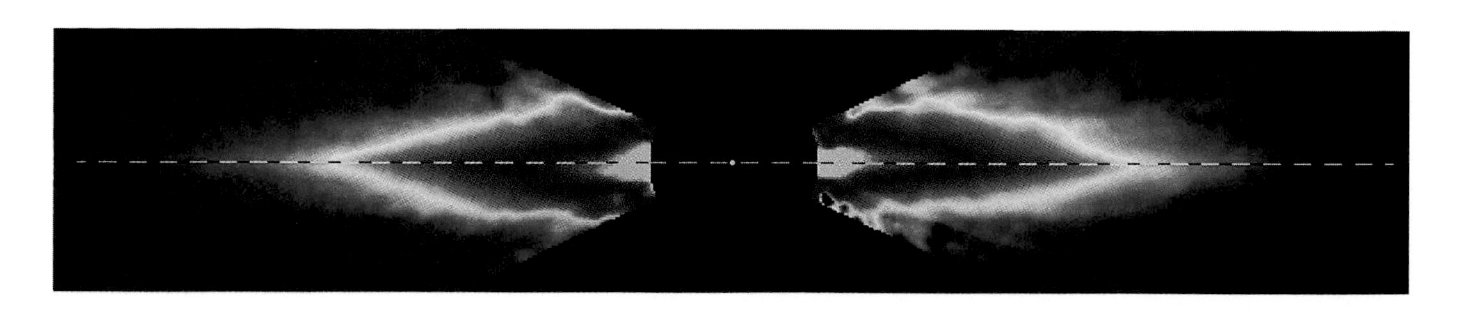

The edge-on dusty disk surrounding Beta Pictoris extends outward for 400 AU from the star, which is covered by an occulting disk to keep it from overwhelming the detector.

ing its capture and the circularization of its orbit. The circumplanetary disks were likely formed differently from the solar nebula, but once established, they developed little "planets," suggesting that planets will form automatically given a source of matter. If that is true, they should form in all circumstellar disks, unless their development is inhibited or disrupted by a binary companion or by the gravitational effect of other stars if the developing system is in a dense group.

Can we find them? Turn the problem around. What would we see of our own system from the nearest star, Alpha Centauri, a mere 4 light years away? Jupiter would shine at 22d magnitude, easily within range of smaller telescopes. It would, however, be at most only 4 seconds of arc away from a brilliant star of magnitude zero that would swamp its feeble glow. Neptune, as it orbits, would twice in a century be 24 seconds of arc away, but much fainter, dimly glowing at 24th magnitude. More distant stars present much greater observational challenges as the planets would be fainter and their orbits angularly smaller. Direct detections are now not possible, although new generations of instruments might soon succeed.

There are other methods of detection, however. It is common to say that the Earth orbits the Sun, or that the Moon orbits the Earth. But in fact two bodies locked in gravitational embrace always orbit *each other* about a common center of mass that lies between them and whose location depends on the ratio of the masses. The Sun has a tiny orbit about the center of mass of the Solar System, dominated by Jupiter. Since Jupiter has a mass 1/1000 that of the Sun, the Sun has an orbital radius about 1/1000 Jupiter's distance from the Sun (5 AU), or about 0.005 AU, a distance equal to the solar radius. As observers on α Centauri watch the Sun move against the distant stellar background on its course around the Galaxy, it must wobble by its own radius, which would subtend 0.003 second of arc, equal to the diameter of a U.S. quarter in New York City seen from Chicago. Observers are approaching this accuracy, but are not there yet.

We could also look for variations in radial velocity. The Sun plies its tiny orbital path at a speed of about 12 m/s (taking 12 years, Jupiter's orbital period, to move through a rough circle of diameter 0.005 AU, or 750,000 km), a velocity four times the current error limit of our best detectors. For either method, we can improve the chance of detection by observing lower-mass stars; moreover, some planets may well be more massive than Jupiter, the combination resulting in larger, faster stellar motions. The radial-velocity method currently has the edge; it also has the advantage of being independent of distance.

Changes in the radial velocity of 51 Pegasi are produced by the gravitational tug of a planet with at least half Jupiter's mass.

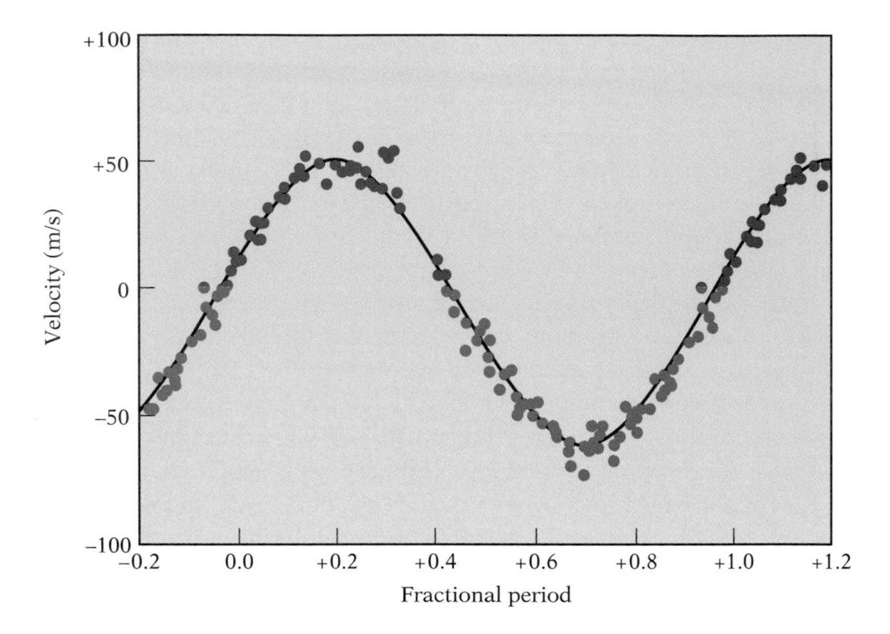

So we look—and there they are. Just to the west of the Great Square of Pegasus lies fourth-magnitude 51 Pegasi, a G2 dwarf like the Sun. Radial-velocity observations show that it has a small, steady, sinusoidal variation that can be produced only by an orbiting companion. Assume that a planetary orbit is circular and that its plane lies in the line of sight. According to Kepler's third law of planetary motion (as generalized by Newton), the period of each body about the center of mass depends on the distance of the planet from the star raised to the 3/2 power and divided by the square root of the sum of the masses of the star and the planet. The planet, by definition, has a small mass compared with the star, so the sum is effectively the mass of the star. The distance between the bodies, however, is effectively the orbital radius of the planet. Since the star lies on the main sequence, we can accurately estimate its mass from its luminosity. Kepler's law thus yields the orbital radius. The maximum stellar radial-velocity deviation depends on the period and the circumference of the star's orbit about the center of mass, from which we find the *star's* orbital radius. The ratio of orbital radii gives the ratio of masses, and thus the planetary mass. However, there is a serious hitch: we have no idea of the orbit's tilt. If, for example, an orbit is tipped perpendicular to the line of sight, we will observe no velocity variations whatever. All we can therefore know are lower limits to the star's orbital velocity, its semimajor axis, and the planetary mass.

Nevertheless, for any reasonably expected tilts, astronomers seem to have found a real planet: the lower limit of 51 Pegasi's companion is only 0.6 Jupiter mass (M_J). A very unlikely almost perpendicular tilt would be required to raise the mass to near that of a brown dwarf. Since that discovery, within a space of less than two years, astronomers found another ten or so stars with roughly Jovian-mass planets and some others with much more massive, yet still apparently substellar, companions. The first set includes a dim, low-mass M star that was analyzed by observing actual positional shifts. The star appears to move in response to *two* planets, each about the mass of Jupiter. The field is exploding, and undoubtedly many more discoveries will be made over the next few years as instrumentation improves.

Most astronomers had naïvely assumed that other planetary systems would be similar to ours. Compared with these systems, however, ours seems hardly typical. The planet that orbits 51 Pegasi is truly weird: its semimajor axis is only 0.05 AU, not much larger than the radius of the star. How a Jovian planet could have formed so close to the star, how it could have survived, or how it could have been dragged in, is a mystery. In fact, with two exceptions, the

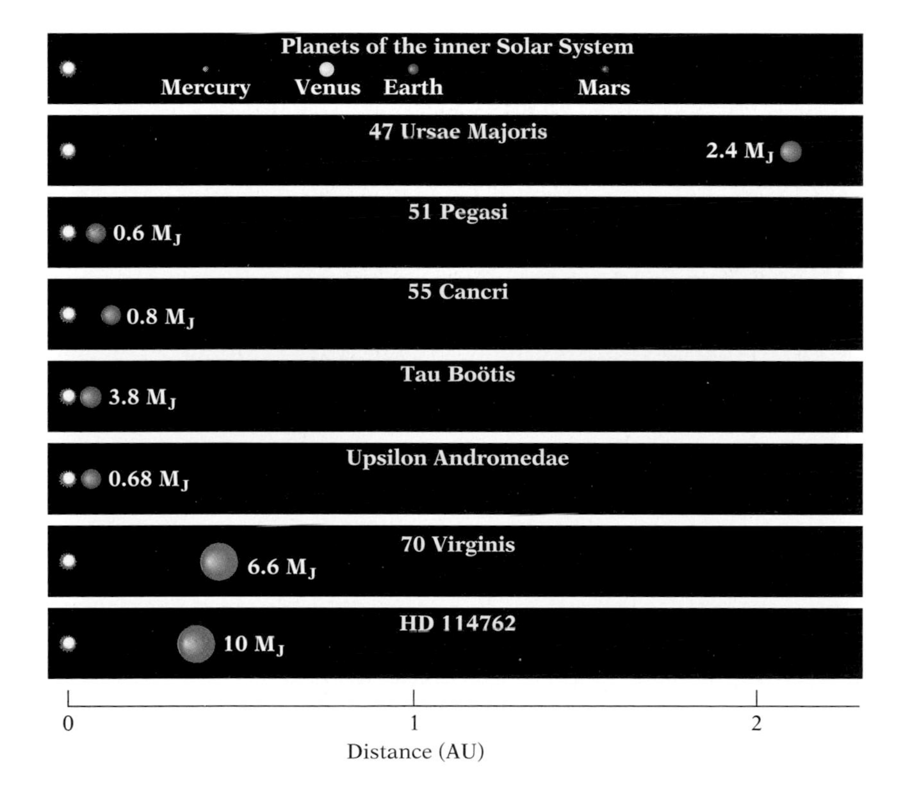

Likely planetary systems, each with only one detected planet, are compared with the inner Solar System; the planets' sizes are greatly exaggerated relative to the parent stars.

systems so far found have Jovian planets in the region where we find our terrestrials. This effect may be caused by observational selection, the perfidious process by which nature first reveals that which is the most obvious, that most amenable to observation. Systems like those found so far would produce the highest stellar velocity changes and would thus be the most likely to be found. Vastly more work needs to be done to determine the distribution of planets; we are, after all, just now *finding* them. But these new discoveries strongly suggest that planetary systems may come in a great variety of forms that depend upon the details of star formation, details that may include the initial cloud's rotation, its mass and mass distribution, its temperature and density, the influences of nearby stars, hosts of other parameters, and quite likely just plain accident. Many of the features of our own Solar System, after all, are accidental in nature, the results of massive occasional and unpredictable collisions.

Actually seeing—imaging—other planets will be much more difficult than indirect detection, as will any kind of detections of terrestrial planets like ours. From α Centauri, Earth would appear at 27th magnitude, near the detection limits of the largest telescopes, only 1 second of arc from the Sun, and would produce a velocity variation in the solar motion only 1/300 that of Jupiter. Imaging at this scale is surely a task for the next century.

But the required technology is near, or just over the horizon. The next generation of *Hubble* instruments, which will eliminate the glare of a bright star, may find extrasolar Jupiters. Vastly more ambitious is a huge multimirror optical interferometer that NASA hopes someday to orbit around the Sun beyond Jupiter, where the local obscuring effects of interplanetary dust are reduced. This major telescopic system, for which much technology has yet to be created, may image planets and, through spectroscopy, teach us something of their atmospheres. One that contains oxygen might signal another Earth.

. . . AND OTHER LIFE

Though there is no way yet of directly imaging other planets, there is another route to detection. Ten percent of the planets of our Solar System—one—harbors intelligent life. Can we assume a similar fraction elsewhere? Our Earth, though optically and dynamically humble, is brilliant in the radio spectrum. Since the days of Marconi we have launched an ever increasing barrage of radio radiation into the cosmos. A bubble of powerful commercial radio transmission now over 60 light-years across encompasses thousands of stars. If someone out there has the same equipment we do, we could easily be found.

Now turn the concept around and look out. The odds of success, first explored by the American radio astronomer Frank Drake in the 1960s through the "Drake equation," seem low in spite of the existence of some 100 billion galactic stars. Assume that double and multiple stars do not form planets. (Which in fact they may well do, the planets either orbiting widely separated components or distantly orbiting the pair. One of the recently discovered Jovian-type planets orbits the secondary star of 16 Cygni.) About 10 percent of all the stars are decidedly single, so the field is narrowed to 10 billion. Computer simulations suggest that circumstellar disks always develop terrestrial planets. To have life, however (reasoning from ourselves), such a planet must fall within a narrow band in which the planetary temperature allows liquid water to be sustained for billions of years. Venus is too close to the Sun, and although we suspect it had oceans at one time, high solar heat made it a runaway greenhouse by evaporating water and baking carbon out of the rocks. If Earth had been much farther from the Sun, it could have become a runaway refrigerator, its water frozen.

No one knows the size of the "living zone"; one study suggested that in the case of the Sun there is only a 10 percent leeway from 1 AU. Moreover, the mass of the planet is highly significant. Mars *did* have liquid water and a sustaining atmosphere but, because of its proximity to Jupiter, turned out too small to keep it. Even if we count Mars, there are still only two possibilities in our system, so, at a guess, cut the number of stars down by another factor of 10. Furthermore, the living zone shrinks dramatically as the mass—and consequently the luminosity—of a main sequence star goes down: the living zone of a dwarf M star is only a fifth or so the size of the Sun's. The odds of a terrestrial planet's falling into a zone that size seem vanishingly small. Seventy percent of the stars in the Galaxy are M dwarfs, so cut the number by another factor of 10. Now we are down to 100 million. The stars so far found with Jovian-type planets have them orbiting close to the living zone, which would probably exclude any terrestrials, so (since we know no better) reduce the number of possibilities to 10 million.

How many of the remaining stars actually develop life? Even if the seeds of life are deposited from interplanetary space via comets and meteorites, it is necessary that the seeds sprout. Locally, we know there is life on Earth and have recently found circumstantial evidence that it may once have existed on Mars. If it did, then we might have reason to assume that life always occurs if the conditions are right, leaving the count at 10 million. But given the not at all unlikely possibility that the Martian discoveries are wrong, then perhaps we should cut the number by another factor of 10 and drop the count to 1 million stars.

But we are not considering just life—ladybugs and grass do not broadcast intelligent radio signals. Life developed on Earth 3 billion years ago, and it took nearly all that time for *human* life to evolve. That time scale removes from consideration all short-lived stars on the main sequence, those from classes F to O. Fortunately for this numbers game, they constitute only 4 percent of all stars, so we can ignore them except to say that Vega, Fomalhaut, and β Pictoris are probably excluded from the club. The Galaxy is about 15 billion years old, so (assuming a constant star-formation rate) two-thirds of its stars have reached the magic number.

Are we a natural by-product of evolution or an accident that may have been dependent on the catastrophe that took out the dinosaurs? An event of that magnitude may have occurred only once since the development of life, a view supported by the ages of the great lunar basins. In the case of Earth, the catastrophe had to happen at the right time, 65 million years ago out of a time span of 4.5 billion, about 1/100 the age of the Earth. So, as a wild guess, cut the number of stars by the same factor to 10,000. And this crude estimate does not factor in how long a civilization might survive. Given the huge error in each successive estimate, we could really choose almost any number we like. There may be millions of civilizations or just one, ours; the various views are more philosophical than scientific.

If other civilizations do in fact exist, how do we communicate? Do we look for bubbles from errant alien "commercial radio"? Do we look for signals meant specifically to contact us, never minding that no one knows we are here? Where do we examine the huge and crowded radio spectrum? Then there is the problem of technical parochialism. Before the invention of radio, futurists proposed contacting the Martians by writing huge letters on Earth, a futile means of communication. Today we have radio—but is radio equally primitive by galactic standards?

Nevertheless, the search for extraterrestrial intelligence—"SETI"—is a scientific endeavor. In 1960, Frank Drake mounted Project Ozma to monitor two nearby Sunlike stars, ϵ Eridani and τ Ceti. He tuned his receiver to a position near 21 cm, since an alien engineer would probably pick the most prominent interstellar emission line as a reference for beginning listeners. Drake found only silence. Several other projects culminated in the High Resolution

Microwave Survey, whose receivers could monitor 20 million narrow-band radio channels at once. Such projects face considerable political problems. NASA helped develop the HRMS but was forced to drop it as a perceived waste of resources (it was immediately picked up by a variety of private groups). The negative results of its successors to date are offset by the huge numbers of stars yet to be examined.

Have we strayed too far from our objective, which was to delineate the natures of interstellar space and the formations of stars and planets? Not only is the Earth a distillation of the interstellar medium, but so are we. Except for hydrogen, all the elements of which we are made—carbon, oxygen, calcium—were created in stars and distributed into interstellar space. We are not just observers of the cosmic scene, but part of it, as natural a part as the stars and planets themselves. However many civilizations there may be in the Galaxy, it still took all of it to make us, and we are as much a part of the study as anything else.

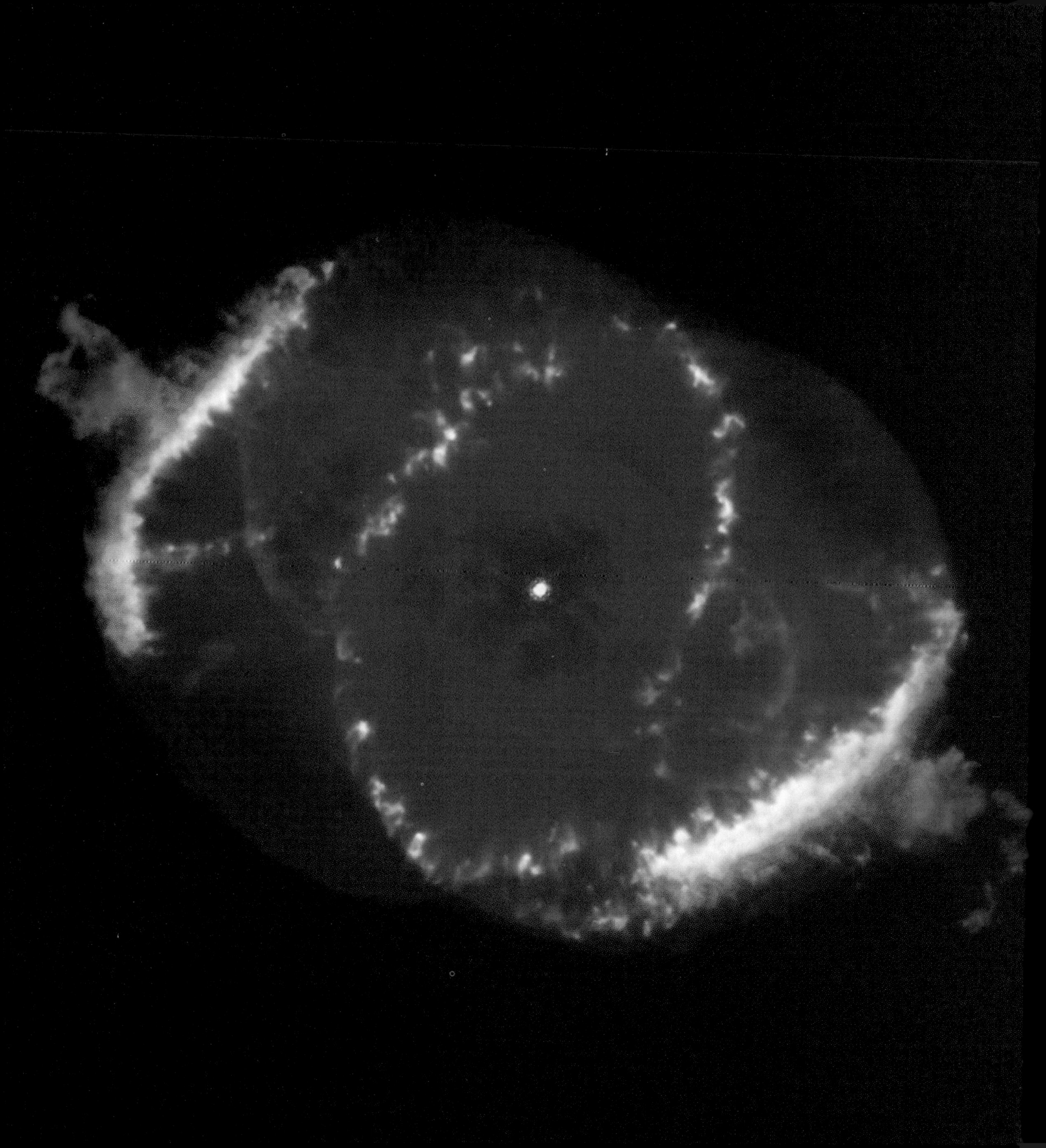

8

PLANETS TO PLANETARIES TO PLANETS

The awesome complexity of NGC 6543 is imaged by the Hubble Space Telescope. Red shows the distribution of hydrogen (from an image taken in the light of Hα), green that of singly ionized nitrogen.

William Herschel, seen here in a painting celebrating his discovery of Uranus, the "Georgian planet."

At the time of the American Revolution, the known Solar System ended at Saturn, just as it had in Homer's day. Its extent doubled in 1781 when William Herschel came across a body "visibly larger than the rest" ("the rest" referring to the stars of constellation Gemini, in which he saw the new body). He initially thought it was a comet. However, its uniform circular motion, Kepler's laws, and its angular diameter of 4 seconds of arc quickly showed that it was no comet but a major planet 19 AU from the Sun and four times the size of Earth. Herschel called it Georgius Sidus ("George's Star") after the monarch of his adopted country, George III. But since that title had little appeal for the French and Germans (or anyone else not English), another name was eventually adopted: that of Uranus, Saturn's mythological father and the personification of the heavens themselves.

Uranus was the trigger for further expansion. Early inadvertent sightings and later observations showed that it refused to behave as predicted under Newtonian gravitational theory. From its orbital deviations, John Couch Adams in England and Urbain Leverrier in France realized the cause of the irregularity might be a perturbing transuranian planet yet unseen. By 1846, after some two years of calculation, they had independently placed the mystery body near the Aquarius-Capricornus border. Johannes Galle in Berlin found Neptune almost immediately after receiving the prediction from Leverrier. (Adams's difficulties in trying to get the English to look for it would have embarrassed Herschel.) Though it now seems that the discovery of Neptune is ancient history, the planet will not complete its first observed orbit until the year 2011.

Even with the effects of Neptune factored in, Uranus did not behave as predicted, so the trick was tried again. In 1929, an American observer, Clyde Tombaugh, was hired at Lowell Observatory to look for *another* perturbing body; in 1930, he came upon Pluto, a body clearly too small to affect Uranus. The problem of Uranus's apparent wanderings was finally solved when astronomers could use accurate masses of Jupiter and Saturn derived from *Voyager* data. But it was Herschel's observation 200 years earlier that was the great leap forward, ultimately extending the planetary system to 30 AU from the Sun and leading to the discovery of what now appears to be the first body of the Kuiper belt—and it was Herschel who launched the study that eventually linked the formation of the planets to the evolution of the stars.

Herschel's Planetaries

In 1785, Herschel wrote the following in the *Philosophical Transactions* of the Royal Society, in an article entitled "On the Construction of the Heavens":

> I shall conclude this paper with an account of a few heavenly bodies, that from their singular appearance leave me almost in doubt where to class them.
>
> The first precedes ν Aquarii 5′·4 in time, and is 1′ more north. . . .

After discussing four more of these mysterious, and nebulous, objects, he continues,

> The planetary appearance of the first two is so remarkable, that we can hardly suppose them to be nebulae; their light is so uniform, as well as vivid, the diameters so small and well defined, as to make it almost improbable they should belong to that species of bodies. On the other hand, the effect of different powers seems to be much against their light's being of a planetary nature, since it preserves its brightness nearly in the same manner as the stars do in similar trials.

With this thoughtful commentary, Herschel began his examination of the "planetary nebulae." He knew that these small lovely

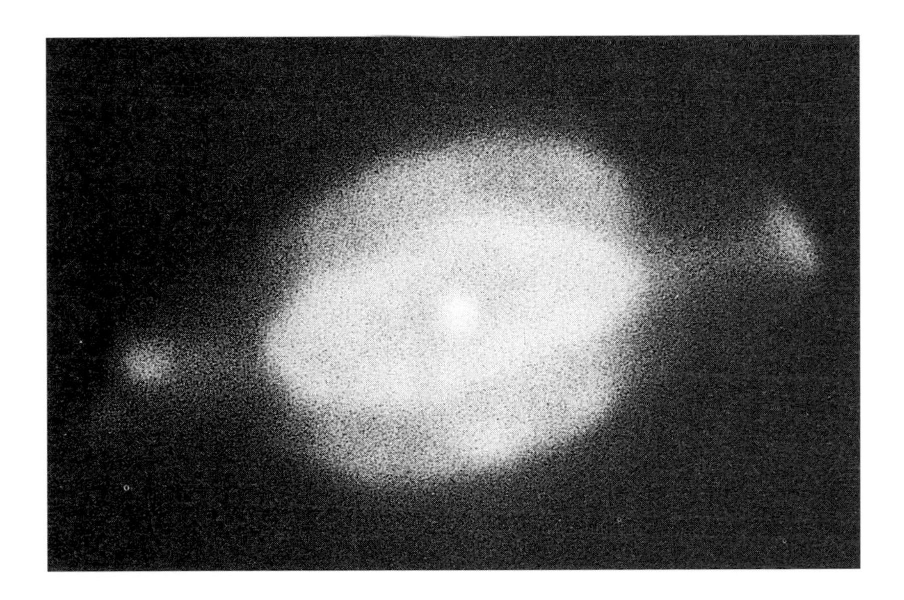

NGC 7009 in Aquarius is the heavenly body for which Herschel originally coined the name "planetary nebula." Its structure includes a smooth ellipsoidal disk from which emerge ansae ("handles") that, in a further irony, give it the popular name "Saturn Nebula." This nebula can be seen, though not as a disk, with binoculars.

nebulae were not planets and had used the term only to describe their disklike appearance. Six years later he described another defining characteristic of these objects:

> On the 15th of February, 1786, I discovered that one of my planetary nebulae had a bright spot in the center, which was more luminous than the rest, and with long attention, a very bright, round, well defined center became visible. I remained not a single moment in doubt, but that the bright center was connected with the rest of the apparent disk.

What was this center? Herschel himself found out. His paper of 1791 reported a series of observations, this one made on November 13, 1790:

> A most singular phaenomenon! A star of about the 8th magnitude, with a faint luminous atmosphere, of a circular form, and of about 3′ in diameter. The star is perfectly in the center, and the atmosphere is so diluted, faint, and equal throughout, that there can be no surmise of its consisting of stars; nor can there be a doubt of the evident connection between the atmosphere and the star.

The picture is now complete: planetary nebulae, nebulous objects that do not themselves consist of stars, have single stars at their centers. The central star was, in Herschel's words, "involved in a shining fluid, of a nature unknown to us." In a striking coincidence, the planetary nebula in which Herschel first saw a central star—NGC 6543—was the very object targeted by William Huggins when he discovered, some 75 years later, emission lines indicating that the "shining fluid" was a gas, thus proving that Herschel was right.

Herschel thought that the star might be forming from the nebula, and upon visiting NGC 6543 Huggins had asked, "Was I not about to look into a secret place of creation?" The direct answer to the question, as we now know, is no: the central star does not condense from the nebula. Indirectly, however, Huggins had expressed a glimmer of truth. Planetary nebulae are not stars being born, but are markers for stars that are dying. Through their deaths they will contribute mightily to the formation of future stellar generations.

Form and Structure

Planetary nebulae—"planetaries"—abound. More than 2000 are known in our Galaxy. Like the stars, they concentrate to the plane of the Milky Way and to the galactic center. The smooth disks seen by

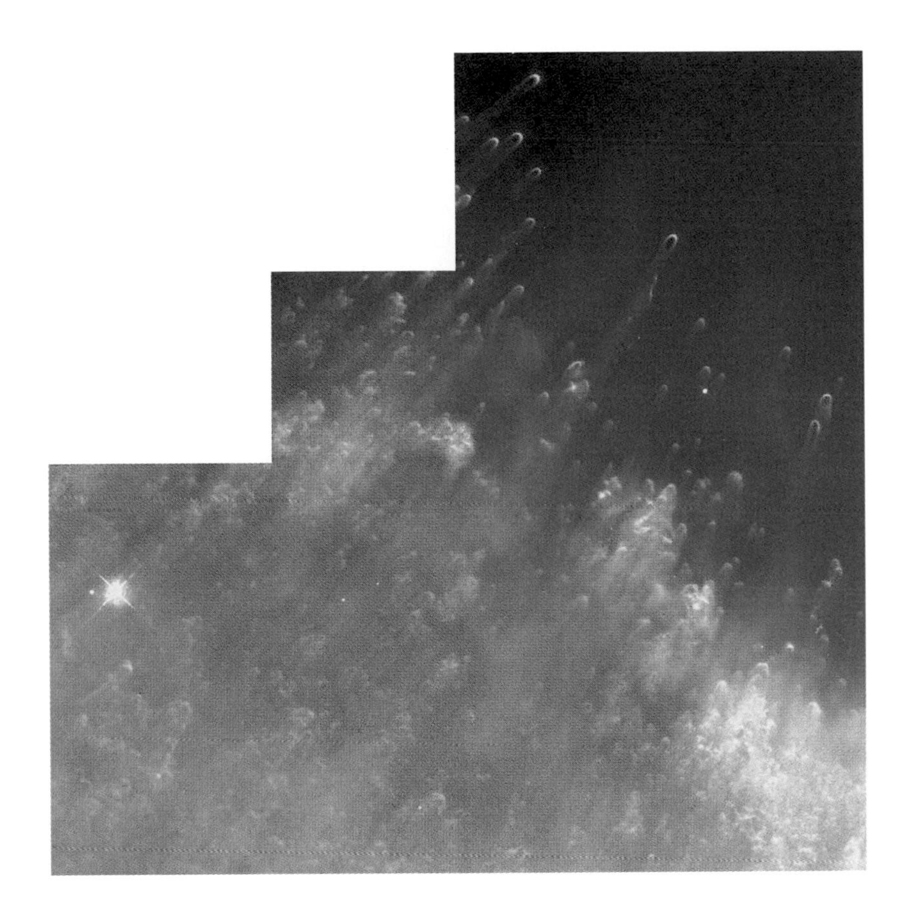

Above, the Helix Nebula (NGC 7293), the closest and one of the loveliest of planetaries, is half the angular diameter of the full Moon. Within its interior are hundreds of bright, dusty, gaseous "droplets." To the left, ultraviolet radiation from the central star ionizes the leading edges of the droplets, resulting in cometlike structures that coalesce into the ring. Each of the droplets is several times the size of our planetary system.

Herschel were an artifact of limited telescopic resolution and poor atmospheric conditions that blurred the forms. When their detail is resolved they provide some of most graceful sights the sky has to offer.

Planetary nebulae range in angular dimension from unresolved pointlike objects that look like stars, recognizable as nebulae only by their emission spectra, to huge constructions nearly the angular diameter of the Moon. To learn their physical dimensions requires knowledge of their distances, and therein lies a massive problem. With a few exceptions the planetaries are too far for parallax measures. Moreover, the spectra of their central stars are quite different from those of the dwarfs, giants, and supergiants of the HR diagram and have not been well calibrated in terms of luminosity. As a result, we cannot determine spectroscopic distances as we can for the exciting stars of diffuse nebulae. Generally we have to turn to a variety of approximate methods that have large errors (for example, the estimation of distance from the degree of interstellar reddening by dust). Nevertheless, it is clear that planetary nebulae range in diameter from only a few thousandths of a parsec to over a parsec and

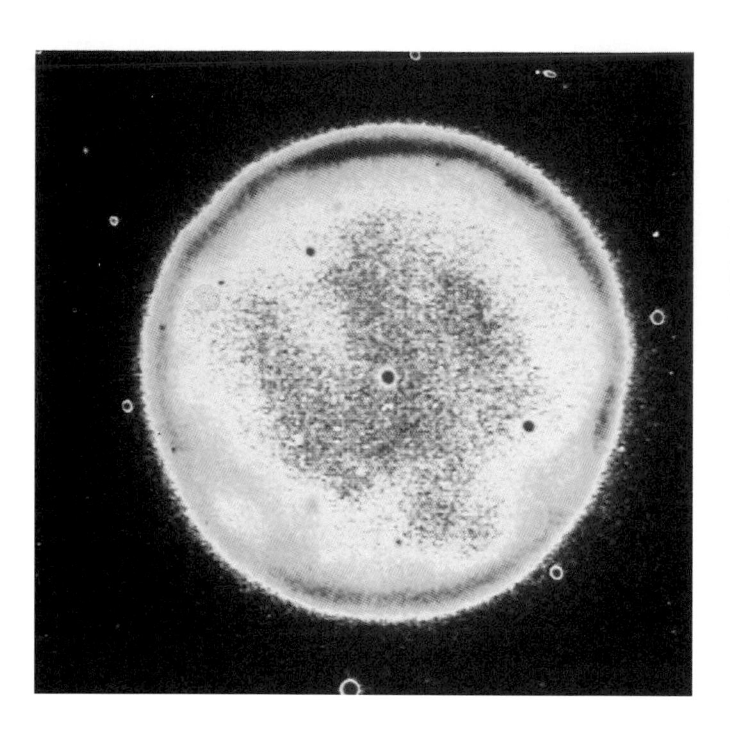

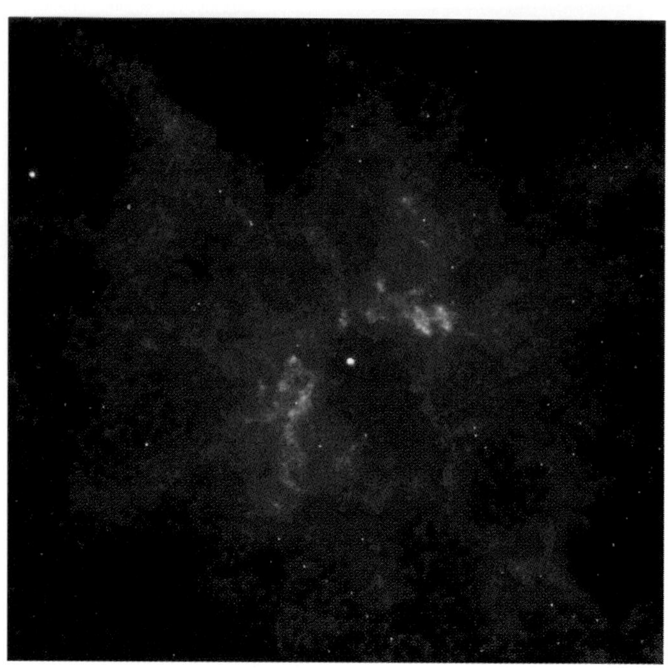

thus can easily span interstellar distances. From the average separation between nebulae, astronomers estimate that the Galaxy contains well over 100,000 of them.

The structures are captivating. Many, like the Helix Nebula (NGC 7293), consist of graceful elliptical rings that are brightest along their minor axes and have a "blown-out" appearance along their major axes. The darkness of the central holes shows that these nebulae are relatively hollow shells. A few are almost perfect rings, while others display a doubled bipolar structure. Quite a large number also have a double-shell appearance, in which a bright ring is surrounded by a fainter one, and many have enormous extended outer halos.

More intriguing details are uncovered by deeper observation. Planetaries are not at all the uniform smooth structures Herschel saw but instead consist of numerous delicate filaments. Inside the Helix are hundreds of tiny gaseous "droplets" with long tails that look vaguely like comets, all pointing away from the central star. More and more of these structures cascade away from the center until they coalesce into the familiar ring. The Helix is one of the nearest nebulae; other planetaries would presumably look much the same if we could see them close up. Several, like NGC 7009, have rodlike "handles" called ansae that project outward along the major axis of the nebula.

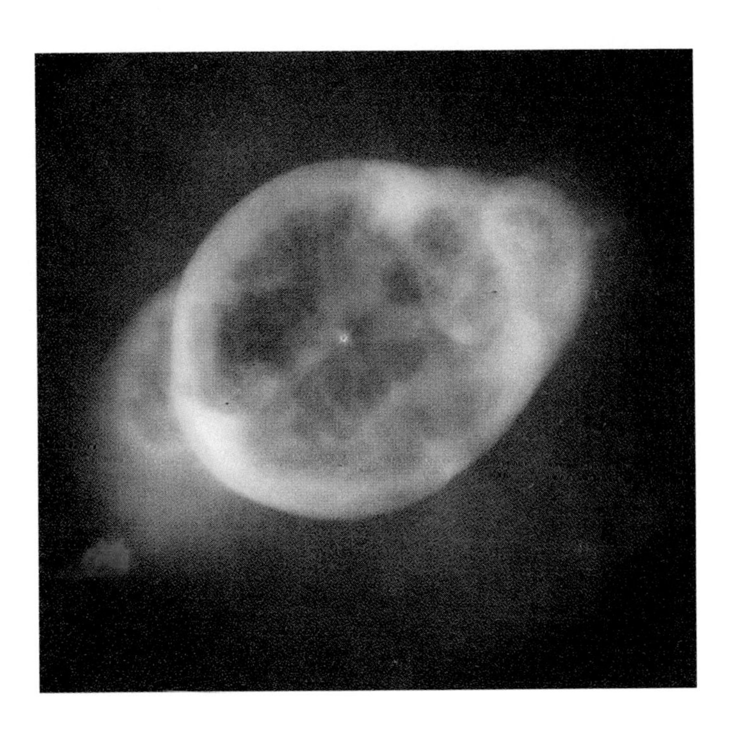

Far left to right: Abell 39 is a perfect ring, NGC 2440 displays a pair of bipolar lobes, and NGC 3242 shows an inner bright ring surrounded by a fainter outer one; Abell 39 is about four times the size of each of the two other nebulae. In the image of NGC 3242, red, green, and blue respectively show the distribution of singly ionized nitrogen, doubly ionized oxygen, and doubly ionized helium.

The awesome complexity of the planetaries is best revealed by none other than Herschel's and Huggins's (and the constellation Draco's) NGC 6543. The first to be seen with its central star and the first to be analyzed by spectroscopy, it was also the first to be imaged (in 1994) with the repaired *Hubble Space Telescope*. Through the extraordinary capacity of this instrument for resolving fine detail, we see narrow double shells (only barely discriminated with Earth-based telescopes) punched through by jets similar to NGC 7009's outer ansae—in fact, a whole spray of jets is visible. J. Patrick Harrington, the astronomer who led the *Hubble* team to take the image and who was faced with the task of constructing a theoretical model of the nebula, was heard to say, "If I had known they were *this* complicated I would have gone into some other line of work." Surrounding the whole affair is the largest and brightest of the outer halos. There is no question that other nebulae imaged with equal sophistication would reveal structures of similar complexity, indeed structures not yet dreamed of. And these are just the optical pictures. Planetaries also shine brightly from the ultraviolet through the optical and infrared to the radio. Imaging alone, however, does not reveal how these beautiful objects are constructed. We need to return to the work begun so long ago by Huggins; as always, spectroscopy provides the key.

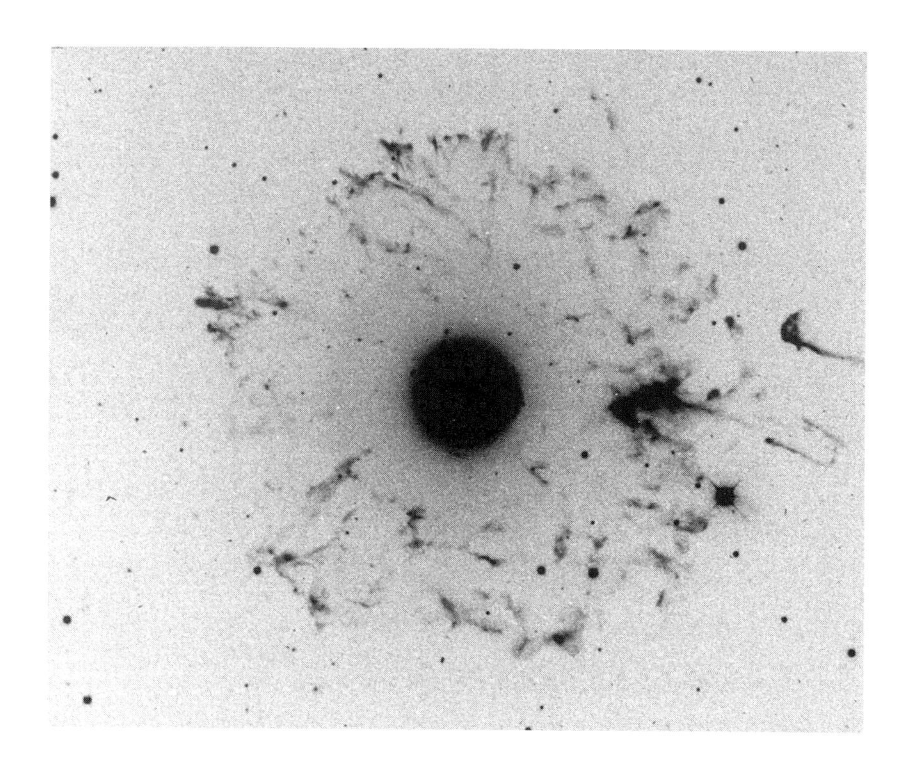

NGC 6543, 0.1 pc across, is surrounded by an enormous and very ragged outer halo. The nebula proper, one-fifteenth the size of the halo, is overexposed at center. The negative image, common in astronomy, brings out faint detail.

Lines, Lines, Lines

At first glance, the spectra of planetary nebulae are similar to those of their bigger relations, the diffuse nebulae. Like them, planetary nebulae shine almost entirely by the radiation of discrete emission lines that belong to particular species of atoms, molecules, or even dust grains. Underlying the emission is a faint optical continuum that becomes strong in the infrared and dominant in the radio. In the optical spectrum are the familiar Balmer series of hydrogen and the same forbidden "nebulium" lines of [O III] (radiated by O^{+2}), [Ne III], [O II], and others. It was, after all, Hermann Zanstra's studies of planetary nebulae that had led to an understanding of the radiation processes of diffuse nebulae. The difference between the spectra of the planetaries and the diffuse nebulae is in their variety: the planetaries exhibit a far greater range of ionization. Though a few bright major objects like the Orion Nebula display strong [O III] lines, most diffuse nebulae are dominated by [O II]. Those of lower ionization have weak recombination lines of neutral helium (He I), which strengthen as [O III] becomes more prominent. However, no diffuse nebula radiates lines of the next stage, *ionized* helium (He II), which is a characteristic signature of the majority of planetaries.

In a few planetary nebulae, the He II line at 4686 Å actually dominates Hβ at 4861 Å. Consistent with the presence of He II are emissions of several other highly ionized species, including lines of [Ar IV], [Ar V], [K V], and at the extreme even [Fe VII], iron with six missing electrons.

Recombination lines, like those of hydrogen and helium, are produced by the capture of electrons by ions of the next higher ionization stage. Although the Balmer lines are radiated by neutral hydrogen, they are created when the ionized atoms (bare protons), which dominate the gas, recombine with electrons. For the nebular gas to radiate He I lines, He^{+} must dominate the helium species, and for it to radiate He II, He^{+2}—fully ionized helium—has to dominate. Hydrogen must absorb a minimum energy of 13.6 eV—the energy of an ultraviolet photon with a wavelength less than 912 Å—for its electron to be kicked away. Helium, with two protons, has a stronger binding force. It takes twice as much energy to lift away one electron (504 Å in the deep ultraviolet) and *four* times as much to rip away the second (228 Å, nearly in the X-ray spectrum).

Diffuse nebulae are illuminated by O stars, the hottest of which have temperatures of about 50,000 K. But a temperature of at least 60,000 K is required to produce photons shortward of 228 Å and thus to create nebular He^{+2}. That the planetary central stars are capable of doubly ionizing helium indicates that they can be much hotter than even the grandest O stars. We are clearly dealing with something most unusual.

At the other spectral extreme, planetaries also display infrared and radio emissions from simple molecules, including H_2, CO, OH, HCN, and HCN^{+}. Their discovery was something of a surprise, since astronomers had expected that molecules would be destroyed in the vicinities of such hot, ionizing stars. The molecules must somehow be shielded from destruction, and the obvious protector is large amounts of embedded dust, which is seen in the infrared by its continuum emission.

The emission spectrum provides a marvelous way of exploring nebular structure. The gas is ionized by the absorption of ultraviolet light from a very hot star. Regions closest to the star generally have the highest ionization levels and regions farthest from the star the lowest. In a typical nebula, He II radiates from the center, where the helium is doubly ionized, and He I comes from farther out; [O III] is more or less centered, and [O II] and [N II]—if there is any—shines near the edge. Since hydrogen has only one ionization stage in the nebula, the Balmer series arises from the whole nebula. If we then take pictures with different optical filters that isolate specific spectrum lines, we can look at different parts of the nebula. In the Ring

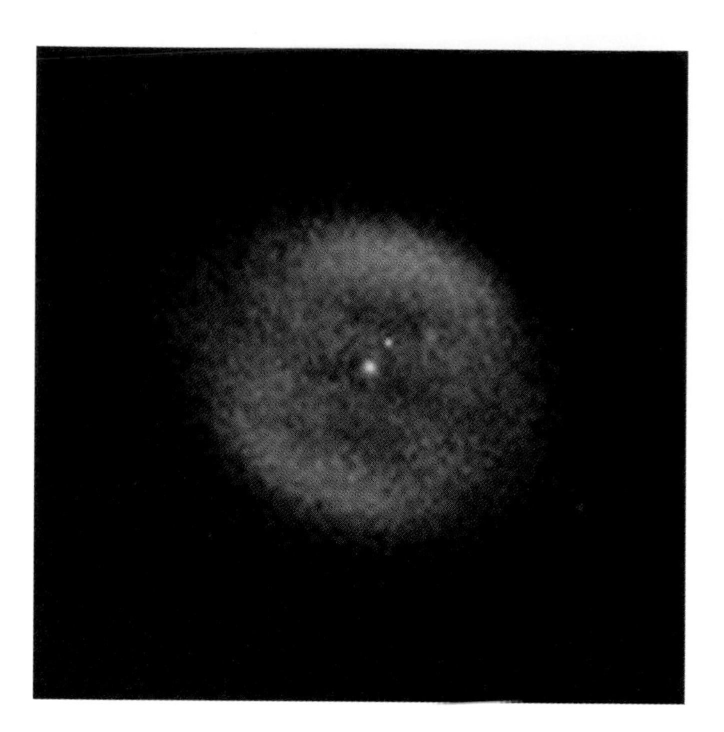

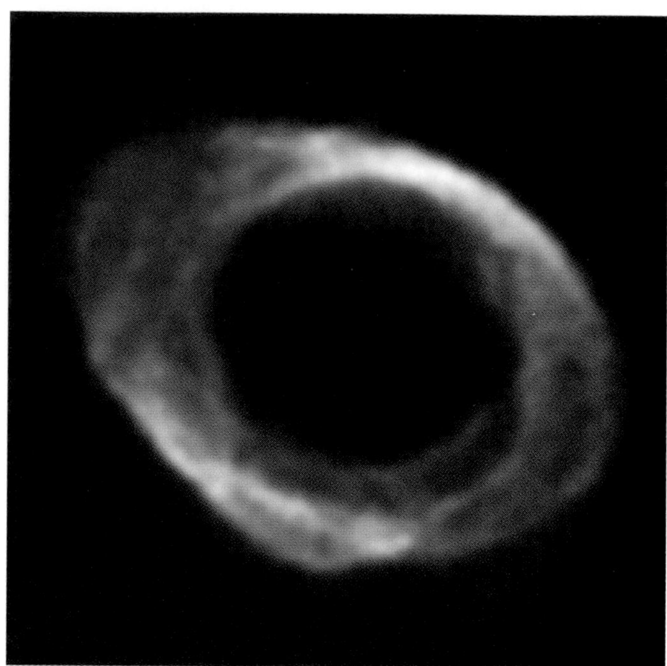

The Ring Nebula in Lyra, seen in images made in the light of two ionization stages: left, He II, and right, lower-energy [N II].

Nebula in Lyra (M 57), for example, He II shows the inner smooth structure, [N II] the outer filaments. The *Hubble* image of NGC 6543 that opens this chapter is a composite, in which red reveals the distribution of hydrogen through Hα and green that of N^+ through [N II].

This technique of spectrum-line imaging can be extended into the infrared and even into the radio. At a wavelength of 3.3 μm lies a spectral feature almost certainly produced by radiation from warm carbon-based dust, perhaps PAHs or just plain soot. Imaging within this emission line shows that in some nebulae the dust (most of which is probably pure carbon) is pervasive, while in others it hovers around the outside, where it can hide portions of a nebula and distort its optical appearance. Imaging in the infrared H_2 lines,

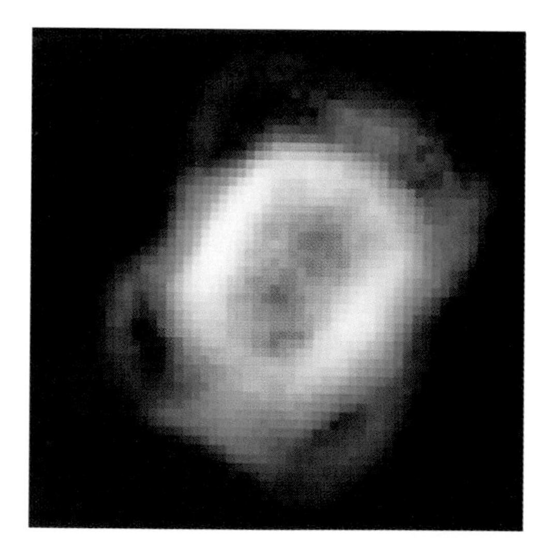

NGC 7027 is one of the most highly ionized and excited planetary nebulae known. In this composite infrared image, the area of distribution of hydrogen is white (through the Brackett α 40.5 μm line), that of dust green (through the characteristic spectral feature of dust at 3.3 μm), and that of molecular hydrogen blue (through a spectrum line at 2.122 μm).

or in the radio CO lines, shows that in some cases the molecules are hidden by the dust within dense, opaque knots of the kind that fill the Helix Nebula. Many other nebulae are optically thick to the central star's ionizing radiation—that is, they absorb it all. These planetaries are thus Strömgren spheres, ionized bubbles within dense outer neutral clouds. All that gets through is lower-energy radiation that slowly breaks apart the molecules that float in the outer reaches, beyond the optically visible parts of the objects.

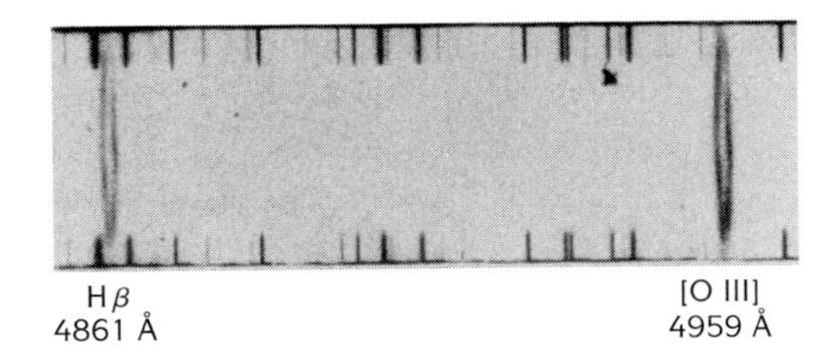

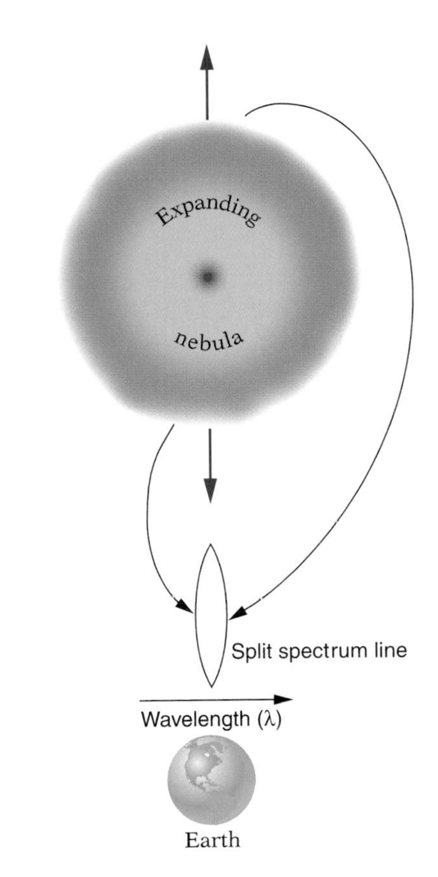

The Hβ and [O III] lines of NGC 6818 are neatly split in two. The front side of the gaseous shell that contains O^+ is coming at us, and the radiated lines at 4861 Å and 4959 Å are Doppler-shifted to shorter wavelengths; the back side is moving away, and its line is shifted in the other direction. The size of the split depends on the expansion velocity.

What It's Like Inside

The observed spread in nebular ionization is matched by a similar range in physical conditions. While the electron temperatures of diffuse nebulae are in the neighborhood of 9000 K, those of the fully ionized regions in planetaries correlate with ionization level and range from 8000 K all the way to 20,000 K. Densities go from the minimum seen for diffuse nebulae, close to 10 electrons (and of course protons) cm^{-3} to nearly 10^6 cm^{-3}, well beyond that seen for ordinary diffuse nebulae. The outer molecule-rich regions, if they are present, are cooler, with temperatures in the hundreds of kelvins.

Ionized masses, derived from the densities and radii, are small. While diffuse nebulae commonly contain the masses of hundreds of suns, the planetaries embrace mere fractions of suns, typically a few tenths of a solar mass. Were we to live in such a nebula, we would barely be aware of it: the heavily diluted glow of the object would be spread all over the sky. The molecular masses range from much less than that of the ionized component to much greater. Chemical compositions, found from the emission-line strengths, are varied. The helium content of the Orion Nebula and others is similar to that found in the Sun. A handful of planetaries, however, have ratios of helium to hydrogen that are more than double solar; nitrogen and carbon abundances are in some instances enhanced by over a factor of 10, nitrogen enrichment loosely correlating with that of helium.

The physical natures of the planetary nebulae become clearer when we examine their spectra at high resolution. The lines are commonly split in two, one component Doppler-shifted to the red, the other to the blue, showing that, relative to the center, one side of the nebula is moving toward us, the other moving away. The nebulae must therefore either be expanding or contracting. Contraction, however, is not a physical possibility, as the large sizes and low masses of the nebulae do not allow them to be gravitationally bound. Typical expansion velocities are in the neighborhood of 20 km/s but they can range upward to nearly 100 km/s. The different expansion velocities along different axes in part explain nebular shapes, as the higher-velocity gases will have moved the farthest since the nebula's birth. No matter what the complexity of the expansion, however, it is fully symmetrical about the lone star in the planetary's middle.

The diffuse nebulae are remnants of star formation: the stars have come from the gas. The planetary nebulae present the reverse: *the gas has come from the star.* The central star of a planetary has partially evaporated, sending its effluvia into space in an expanding shell. The odd chemical compositions of some of the nebulae, the enrichments of helium, nitrogen, and carbon, thus make sense: we must be seeing by-products of thermonuclear fusion, of the processes that have kept the star alight for its lifetime. From the expansion velocities and radii we find that even the largest planetaries are only a few tens of thousands of years old. Planetaries are fleeting objects, in the context of stellar astronomy gone in the blink of an eye. To understand the origins of the nebulae, we must have a closer look at the natures of the central stars themselves.

At the Heart of It All

The central stars of planetary nebulae are all blue and hot, as they must be to produce the observed nebular ionization and emission-line spectra. They are not, however, just hotter kin of the O stars that illuminate diffuse nebulae. Instead of the absorption lines normally found, many planetary central stars radiate powerful and broad emission lines of carbon, helium, and oxygen. Others have almost no lines at all.

To explain the illumination of both planetary and diffuse nebulae, Zanstra showed that in an optically thick nebula—a Strömgren sphere that absorbs all the ionizing ultraviolet photons—every ionizing photon is ultimately converted into a Balmer photon. The rules of quantum mechanics give us the rates at which electrons recombine onto different energy levels. As a result, we can easily calculate the ratios of the strengths of Balmer lines and relate the total number of Balmer photons to the number radiated in any single Balmer line; Hβ at 4861 Å is traditionally taken as the standard. The number of Hβ photons therefore tells us the number of ionizing ultraviolet photons, those shortward of 912 Å, that are radiated by the star.

Look at what can be learned from Zanstra's insights. The apparent magnitude of the star provides the number of photons arriving at the Earth per second in the yellow part of the spectrum. The total number of ionizing stellar photons (those shortward of the ionizing limit of 912 Å), the number that *would* be arriving at the Earth were the nebula not intercepting them, is determined by the number of arriving Hβ photons. Since the exciting star is assumed to be a blackbody, the ratio of the number of ionizing photons to yellow visual photons depends critically on the star's temperature, which can then easily be derived. If the nebula is optically thin—that is, if ionizing photons reach the edge of the expanding nebula and escape into space—all the ionizing photons cannot be counted and thus only a lower limit to temperature can be derived. But the same trick can be performed with the nebular He II lines, counting the ionizing photons in the spectral realm shortward of 228 Å, where the nebulae are almost always optically thick to radiation. Zanstra saw the nebula as "an ultraviolet photon counter" and referred to the mechanism as "space research at low cost."

The "Zanstra temperatures" so calculated for the central stars range from about 28,000 K to 250,000 K, five times hotter than the hottest O star and over forty times hotter than the Sun. The stars that illuminate the planetary nebulae are (with the possible exception of the tiny remnants of exploded stars) the hottest in the Universe. Though they appear dim to us—none is even close to being visible to the naked eye—they are still among the most luminous of stars, shining thousands of times brighter than the Sun. They appear faint only because of their distances and because most of their radiation—90 percent or more—flows outward in the invisible ultraviolet.

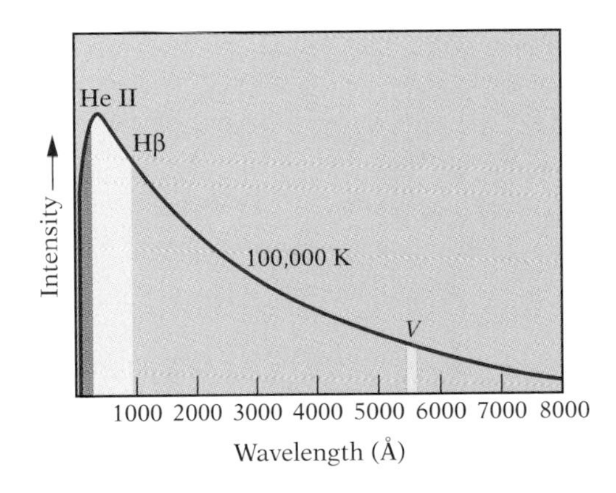

The continuous spectrum of a blackbody at 100,000 K is a good approximation to that of a planetary nebula nucleus at the same temperature. The number of yellow photons arriving at the Earth is given by the star's visual magnitude (V). The number of ultraviolet photons shortward of 912 Å and 228 Å are given by the strengths of nebular Hβ and He II, respectively. The ratio of ultraviolet to optical photons depends very strongly on temperature.

The central stars link directly with the expanding planetary shells. The emission lines often seen in optical spectra and the powerful P Cygni lines (emissions flanked to the short-wave side by absorptions) commonly visible in the ultraviolet are the signatures of strong continuing mass loss, of stellar winds with rates of around a billionth of solar mass a year or more, 10,000 times the flow rate of the solar wind. More surprising are the winds' speeds, which range from a few hundred kilometers per second—comparable to that of the solar wind—to several thousand. These winds must react with the expanding nebulae, and therein lies the clue to the formation of planetary nebulae.

Until the 1960s planetary nebulae were thought to be bizarre, pathological phenomena that had little to do with mainstream stellar astronomy. However, for there to be so many of them despite their short lifetimes, most stars must go through the stage, producing planetary nebulae and losing a good part of themselves back into space. The theory of stellar evolution finally showed how this happens and gave rank to the planetary nebulae as the final active episodes in stellar lives, the last gasps of the stars as they toss their matter back into interstellar space, mass from which new stars will someday be born.

THE COURSE OF LIFE

As soon as we are born, we begin to die; so too with the stars. As soon as they emerge from the interstellar medium they begin the processes that lead them to their ends. Everything that lives must be fueled. For us, fuel comes from outside ourselves; stellar fuel, however, is internal. There is only so much of it, and when it runs out, the star expires. Stars run off two kinds of fuel, gravitational and nuclear, that play both with and against each other. As stars gravitationally contract out of the dusty gases of interstellar clouds, they heat. When the temperature in the deep core of a forming star—perhaps a T Tauri star like our Sun in its youth—becomes high enough, hydrogen begins to fuse to helium by the proton–proton chain. The new energy source stops the contraction, holding off gravity and its now latent energy, and a new main sequence star—as once upon a time did our Sun—comes to life.

Only the inner half of the Sun's mass is hot enough to sustain fusion. As long as there is any hydrogen left there at all, the Sun will remain quite stable, changing only very slowly over the next 5 billion years. Greater masses higher up the main sequence produce higher internal temperatures, higher nuclear burning rates, and dramatically shorter main sequence lifetimes. But whatever the mass, the fuel must eventually run out, and once it does the core resumes its contraction under gravity's implacable force. But now an odd thing happens that is a prelude to the planetaries. As the core shrinks it gets hotter, and hydrogen is fired in a small shell surrounding the defunct core. This new nuclear burning and the release of gravitational energy make a dying solar-type star much *brighter.* As the core shrinks toward the size of Earth, to densities of nearly a million grams/cm^3, the outer envelope, which is not involved in thermonuclear burning, expands. The outer limits of the Sun will reach beyond the orbit of Mercury almost to that of Venus. Enjoy Mercury now: it will be gone in less than 6 billion years. The expansion will cause our aging star to cool and redden, and perhaps someone, somewhere, will see the Sun bloom into a red giant. The large size and resulting low surface gravity, added to the great luminosity, promote a much stronger wind, one that blows 100,000 times more strongly than that from the current Sun. The star is slowly evaporating, the enhanced wind beginning to return serious mass to interstellar space.

But the nuclear engine is not yet done: its companion, gravity, brings it back to life. As the core shrinks, and the helium created by hydrogen burning reaches a temperature of 100 million K, it begins to fuse to carbon (when three He nuclei collide nearly simultaneously) and even to oxygen (as the carbon picks up another helium nucleus). The energy so generated brings the contraction to a halt and the star stabilizes as a red giant—in the case of a solar-mass star, for about a billion years. Inevitably, however, even the helium runs out, leaving the star with a burned-out carbon–oxygen core. Gravity now regains the upper hand, and compression resumes. Increasing temperatures force the helium burning outward into a shell surrounding the core and make the hydrogen burning spread out even farther. And now the star blossoms again, becoming even brighter, larger, and redder than before, a giant of magnificent proportions. On the HR diagram, these stars fall along a locus that is roughly asymptotic to the zone occupied by the earlier giants, and they are therefore called "asymptotic giant branch," or "AGB," stars. The AGB sun will probably reach the current orbit of the Earth. First goodbye Mercury, then most likely farewell Venus.

Wonderful things continue to happen at an ever increasing pace. As AGB stars brighten, they become unstable and begin to pulsate, changing their luminosities, surface temperatures, and radii. There is nothing subtle about such stars: the first was found 400 years ago by David Fabricius. He noted a third-magnitude star in Cetus where he had seen none before. Then it disappeared, and a year later it *came back*. Appropriately, he named it Mira, "the amazing one." As it varies between third and tenth magnitude over a 330-day period, it changes its radius from the size of the orbit of the Earth to about that of Mars. At their peaks, a dozen or more other Mira-type variables are easily seen with the naked eye. Thousands are known, so bright they can be observed a good part of the way across the Galaxy.

Mira pulsations may seem slow—their periods can be up to several years—but the stars are so enormous that the expansions are actually quite vigorous and generate shock waves, sonic booms, revealed by emission lines in the stellar spectra. The shocks appear to help lift gases away from the stellar surfaces at speeds of 10 to 15 km/s. As the gas chills, some of it condenses into dust grains. Radiation pressure from such an enormously luminous star drives the dust outward (so goes the most popular theory), the dust drags the gas, and the result is a fierce wind that can blow with a mass-loss rate of 10^{-5} solar mass a year. Given that AGB stars are in this condition for tens, even hundreds, of thousands of years, they lose a good part of themselves—ultimately the entire stellar envelope that

When the hydrogen in the core of a star is exhausted through nuclear burning, the star brightens and reddens as it climbs the red giant branch of the HR diagram (here plotted with luminosity against temperature). When helium is fired up in the core, the star moves back down the red giant branch and stabilizes. When the helium is burned to carbon and oxygen, the star again climbs the AGB. When, through mass loss, the core is nearly exposed, the star moves to the left, and when it hits 28,000 K or so illuminates its developing planetary nebula. At about 100,000 K, nuclear burning shuts down; the nebula continues to expand, and the core cools and dims to become a white dwarf.

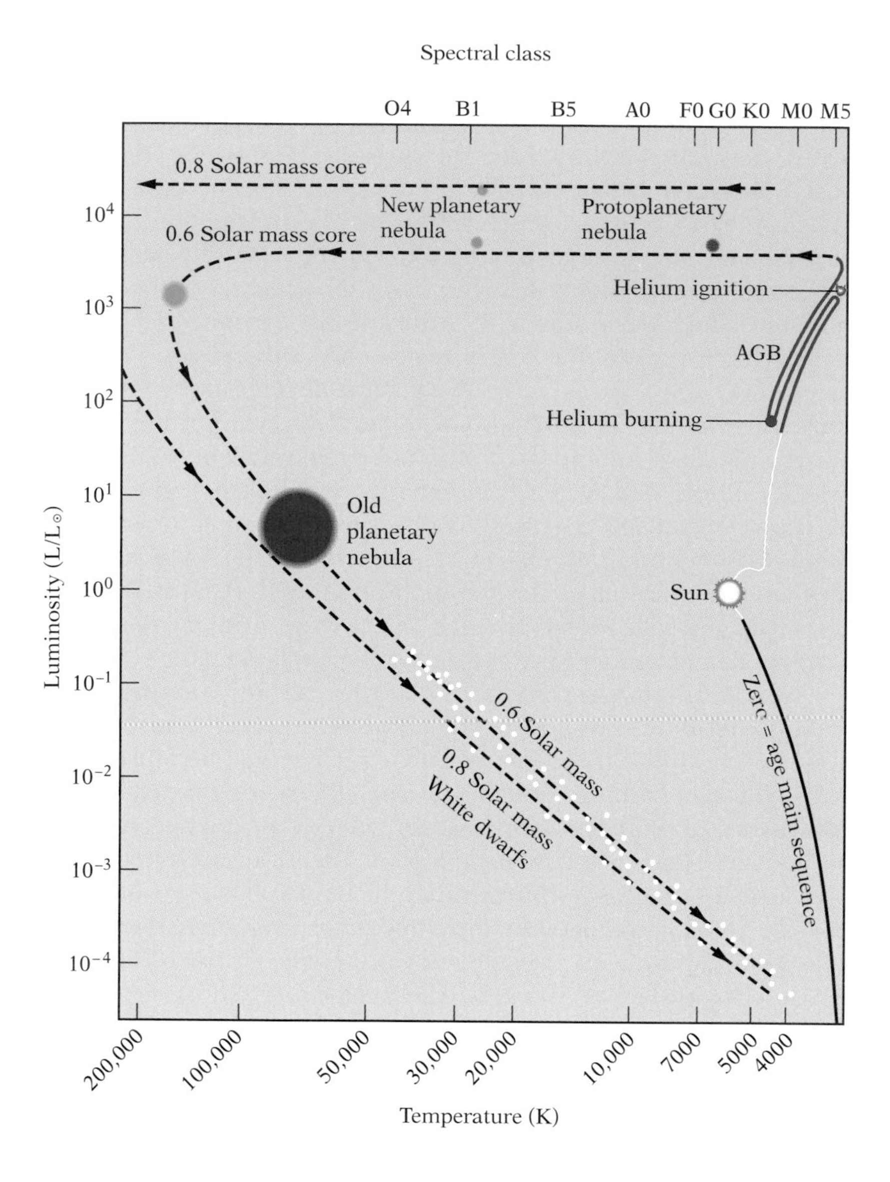

surrounds the dying core—back to space. The winds mean that Earth is expected to survive the solar onslaught that will destroy Mercury and probably Venus: as the AGB sun loses mass, it will loosen its gravitational grip on the Earth, sending our planet spiraling away, out of solar reach (certainly a Pyrrhic consolation since no one could ever survive the intense heat). If mass loss is quick enough at the right time, maybe even Venus can escape.

Most stars have layers in which matter is in convective motion. Where these layers lie depends on the star's temperature and density structure, and thus on the stellar mass and the star's evolutionary state. The outer third of the Sun is in such a convective state. Some AGB stars develop powerful convective zones that can bring matter all the way from near the core to the stellar surface; at the same time, convection within the nuclear-burning shells can bring freshly made elements upward.

If conditions are right the inner and outer convective zones can connect. Newly minted helium, nitrogen, and carbon can thereby be lifted to the stellar surface, in quantities such that from Earth we might—could we live long enough—watch the star change the chemical composition of its outer, visible layers. Most AGB stars have oxygen-rich outer layers, with (typically) three times as much oxygen as carbon. In some, however, including many Miras, the proportions are wildly reversed, the outer layers having far more carbon than oxygen. And along with the carbon comes a variety of other elements created in the inner nuclear furnace by a complex network of nuclear reactions. The capture of neutrons by iron nuclei, for example, can send the nuclei climbing the periodic table through zirconium and technetium (atomic number 43, so radioactive that none appears on Earth) and on to heavier elements. In the right circumstances, much of this stuff can find its way to the top.

The dust surrounding the more massive Miras can become so thick that it literally buries the star, and all we see is the continuous infrared spectrum of warm dust. We observe the infrared signatures of silicate dust around the oxygen-rich Miras, those of carbon-based dust around the carbon-rich stars. Within these thick clouds molecules can form. The oxygen-rich envelopes are loaded with the hydroxyl radical, OH, which produces a natural maser that lies in a shell with a radius of 1000 AU or so. Inside this shell are water

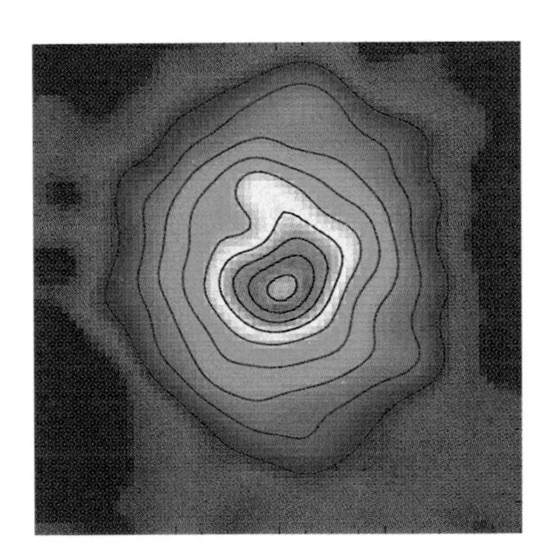

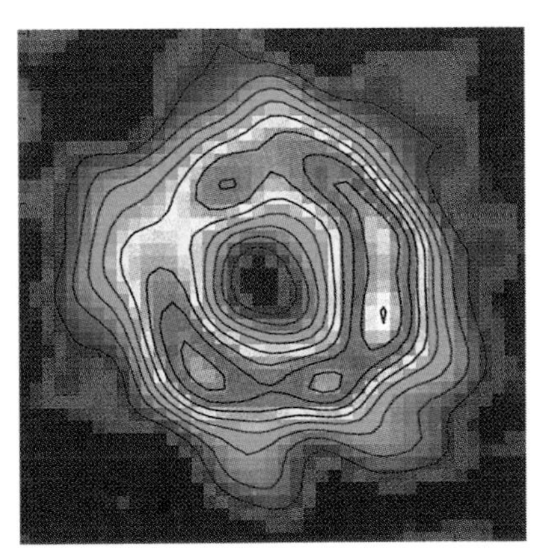

The thick cloud surrounding the carbon star IRC +10216 is rich with organic molecules, as shown by these two radio maps that isolate HCN (left) and HC_3N (right). Each map depicts an area 6000 AU across. The layering is caused by complex reactions among dozens of different kinds of molecules, each of which has different dependencies on density, temperature, and the intensity of stellar radiation.

masers, and close to the star are masers produced by SiO, silicon monoxide. The shielding envelopes around carbon stars are yet more fascinating, as the carbon creates large numbers of complex circumstellar molecules. We see several species of the sort found in molecular clouds, including those of long chain molecules, but also some (including acetylene, ethylene, and table salt) that are the result of the different temperatures and densities of the circumstellar environment.

The Last Gasp

For a billion years or so, an evolving solar-type star loses an increasing amount of mass, the cool red giants and even cooler and redder AGB stars surrounded by ever thicker, nested, expanding clouds of dirty gas. At last, the star's outer envelope is nearly gone, leaving only a relatively thin skin to protect the fury of the nuclear-burning shells from the cold of outer space. The star's apparent surface now shrinks and heats as a result of winds that remove mass from the surface and nuclear burning that eats away at the thin envelope from the inside. On the HR diagram, the stellar cinder suddenly executes a hard left turn as it leaves the tip of the asymptotic giant branch and begins to heat at constant luminosity, changing its spectral class as it goes. The star and its ejecta are preparing a planetary nebula, one perhaps to be seen and pondered by some future galactic Herschel.

The transition from AGB star to planetary nebula is swift and clouded, not just in dust but, because of the dust, by scientific mystery as well. Numerous protoplanetary nebulae—planetaries in the making—are suspected, some even confirmed, but what goes on inside them is still deeply obscure. (The term "protoplanetary nebulae" continues Herschel's legacy of semantic confusion: it is also used to describe disks around *young* stars that are probably producing *planets!*) The complexity of this stage of evolution is beautifully illustrated by the Egg Nebula in Cygnus. The *Hubble Space Telescope* reveals a dark disk from which emerge opposing double searchlight beams set within two dozen circular rings. Ground-based observations had already shown the object to be a reflection nebula with the spectrum of a warm—7000 K or so—yellow-white F star that must be hidden in the dusty center. The rings almost certainly represent numerous episodes of enhanced mass loss. From the expansion velocity of the nebula (about 10 km/s), the events take place every century or so. The beams appear to be produced by ring-shaped polar holes in the dark disk, perhaps punched through by

The protoplanetary Egg Nebula consists of a warm F giant star buried in a thick dusty disk of its own making surrounded by concentric rings punched through by pairs of double "searchlight beams."

some kind of bipolar flow; or they could just be reflection from the thick edges of open polar holes. Molecular imaging shows polar flows about twice as wide.

As the tattered remains of the star's hydrogen envelope dissolve, what is left continues to shrink and heat. To escape the increased gravitational attraction of the smaller core, the star's wind must blow faster, accelerating from the giant's modest 10- to 20-km/s wind to speeds in the hundreds of kilometers per second. This fast thin wind encounters the slower-moving thick wind released by the AGB star, creating a shock wave and shoveling the dusty gas ahead of it into a thick shell. When the stellar surface hits 28,000 K or so, it produces enough ultraviolet radiation to ionize the expanding compressed shell, and there—at the beginning of its ephemeral splendor—appears the planetary nebula.

At first the nebula is so dense that the plowed shell is optically thick, capturing all the ultraviolet photons emitted by the central star. But the little nebula, still only a few hundredths of a parsec across, is expanding. The density drops and the photons penetrate

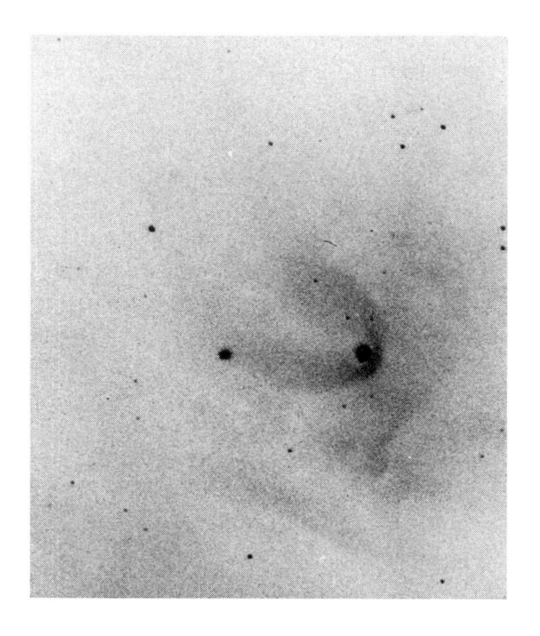

Abell 35 has "hit the wall": the interstellar medium stops and absorbs the expanding shell while the central star escapes, leaving a shock wave behind it.

farther into the cloud. Moreover, the star is heating, the number of UV photons rapidly increasing. As a result, the exciting UV radiation finally reaches the edge of the inner shoveled shell and escapes to illuminate the outer uncompressed region. Now we see a typical double-shell nebula, consisting, as it were, of a skating area cleared of its snow, surrounded by a ring of mounded snow and then the untouched snow-covered surface of the frozen pond. Further expansion can even allow the photons to penetrate the uncleared region and illuminate the early AGB wind to produce outer halos like the ones surrounding our old friend NGC 6543.

When the Sun is in the planetary-nebula state it is expected to be reduced to nearly half its present mass and to reach a surface temperature of over 100,000 K. Earth may escape destruction, but it will be thoroughly fried by its dying star's ultraviolet radiation. Higher-mass stars lose much more matter, leave higher-mass remnants behind, and become substantially hotter, over 200,000 K. In time, however, the outer stellar skin is almost gone, nuclear burning shuts down, and the core cools and dims. At this stage, we begin to see huge, faint, expanding nebulae surrounding dimming stars. The nebula finally starts pushing the interstellar medium in front of it, and the expansion slows and brakes to a halt. The star, however, keeps moving in its galactic orbit and begins to escape its jettisoned layers, which gradually but inexorably merge with the dusty interstellar medium to create raw material for new stars.

The planetary is the interface between the dying star and the interstellar medium. In a sense it is a "transparent star" in which the outer layers are lost to the star and thin out, allowing the inner, old, nuclear-burning core to be exposed. The nebula thus gives us valuable clues about the evolution of giant stars. Its chemical composition, for example, tells us something about how the star cycled its mass from its nuclear-burning interior to its windy surface.

The nebulae are also tattletales about how AGB stars lose mass, as they are exaggerated memories of the original stellar winds. We can explain the majority of the various structures of planetary nebulae by assuming that AGB stars have strong predispositions toward losing mass in their equatorial planes and that there is a huge range in the ratio of equatorial to polar ejecta. If the initial mass loss is almost perfectly symmetrical, we ultimately find a beautiful ring. However, if there has been even a modest enhancement of matter in the AGB star's equatorial plane, the hot fast wind (which blows once the outer atmosphere is sloughed away) expands faster in the thinner regions outward along the poles. As a result, a typical planetary takes on an ellipsoidal structure with the long axis lined up along the star's rotation axis, as we see for Herschel's NGC 7009. The plowed gas eventually breaks through the outer part of the old wind along the ellipse's major axis. At the same time, the ionization front is also expanding. As a result, we eventually see a mature nebula with an equatorial ring set perpendicular to pairs of expanding blown-out bubbles. Finally, if the equatorial-to-polar ratio in the AGB wind is high, we observe a great exaggeration of this picture, the resulting mature nebula displaying a very thick, disklike ring—almost a bar—with much larger bubbles. Different viewing perspectives can give nearly the full range of observed planetary morphologies.

The knots of dark matter seen so vividly in the inner part of the Helix Nebula suggest that AGB wind is filled with inhomogeneities. As the central star's radiation ate into the neutral zone, or perhaps as the hot fast wind plowed up the slow wind, the boundary became unstable, leaving behind the denser neutral knots of dusty matter. Once the central star became sufficiently hot, it began to ionize the knots' outer layers on the starward side, eventually producing the cometlike droplet structures that blended together to make the nebula's ring.

However, even though we might now be able to explain some of the structural properties of planetary nebulae, we have only pushed the mystery back a little further. Why do stars eject mass differently along different directions? The rotation of the originating giant star may certainly be influential, as mass would be removed more easily from the faster-moving equator; magnetic fields promoted by rotation could also play a role. One intriguing possibility is that AGB stars are spun up, as they are during their giant stage, by binary companions. In this picture, the planetaries with extreme bipolar structure would have relatively massive stellar companions, main sequence stars (or white dwarfs, end products of evolution) that still have less mass than the evolving star. AGB stars with mild equatorial wind enhancements, which produce the elliptical nebulae, could be made to spin faster by low-mass orbiting bodies, by brown dwarfs, or even by Jupiter-sized planets. Much of the wind structure could also be explained by a binary companion that is stirring and directing the outflowing gas. Indeed, one school of thought suggests that almost all the different nebular structures could be produced by the gravitational actions of binary companions. Evidence is supplied by complex objects in which the gas coming out in one direction is a mirror image of that coming out in the other, implying wobbles in the ejecting star produced by binary action.

More puzzling are the complex ansae seen in so many elliptical nebulae. From their spectra, the ansae are illuminated not so much

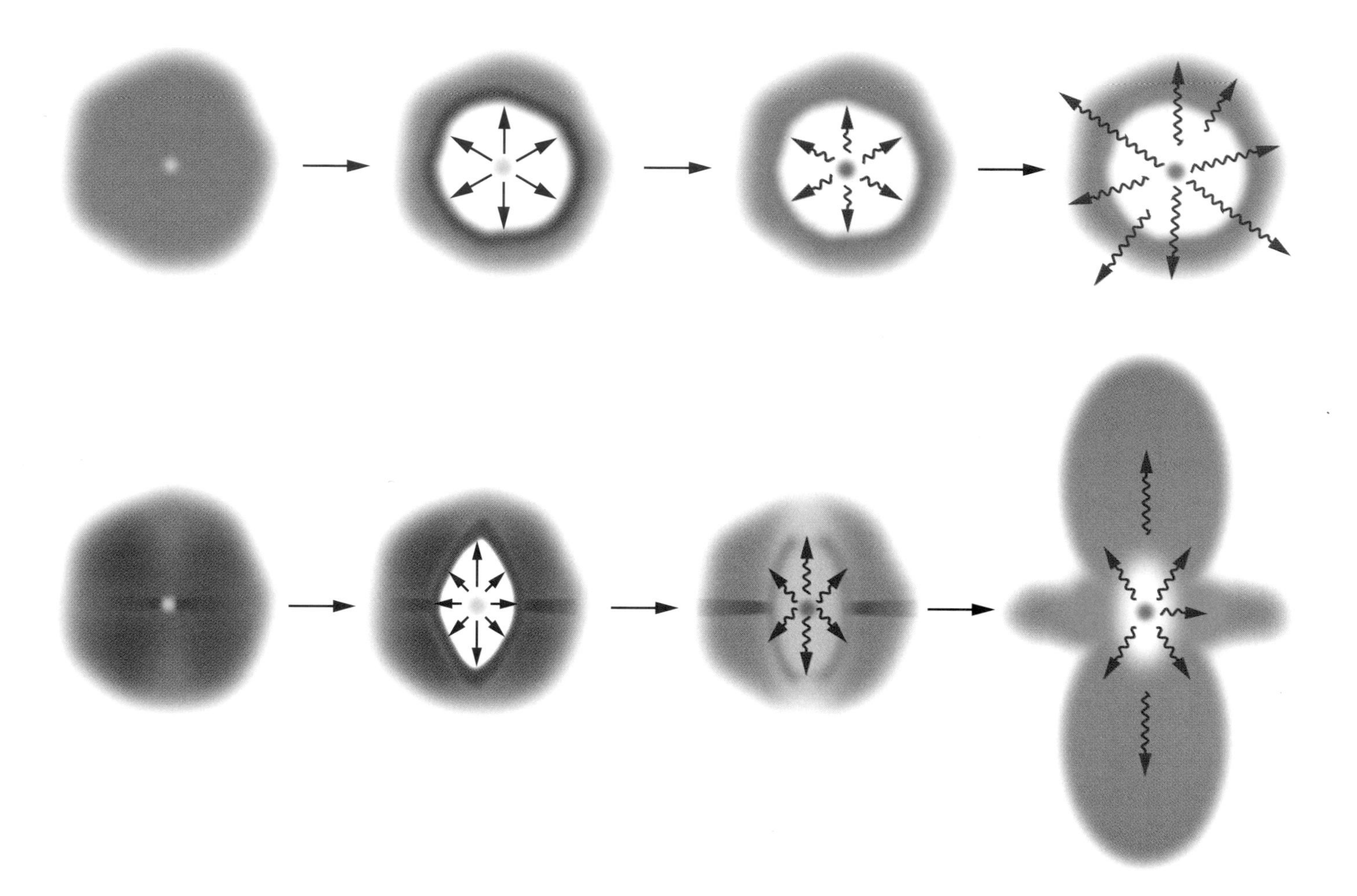

A planetary nebula is produced when mass lost from a giant star (gray) is hit by a fast, hot wind from its nearly exposed core and is shoveled into a dense shell. When the star is sufficiently hot, it lights the shell (green), and when the radiation escapes, it lights the outer, unshoveled portion (pink). The shapes of planetaries depend on their ages and the mass distribution. At the top, mass is spherically distributed, and the nebula is a ring. At bottom, most of the giant's mass is lost at the equator, and the nebula takes on a bipolar appearance.

from ionization by the central star as by shocks, and they must therefore be produced by high-speed bullets of gas ejected by the stars, presumably along their polar axes. Their similarity to Herbig–Haro objects demonstrates a remarkable physical relation between forming and dying stars, which is only enhanced by the cometary droplets seen in the Helix that are reminiscent of the evaporating stellar cocoons seen by *Hubble* in the Eagle Nebula. Are these neo-HH objects also the result of binary action, or of some obscure phenomenon that is not yet at all understood? And why should NGC 6543 display complex sprays of jets? We have no idea at all. To quote Bruce Balick, one of the originators of the morphological concepts presented here, "What are the ansae? How are they ejected? Only the central star knows for sure"—questions that could well apply to the original wind in general.

LIFE AFTER DEATH

The glowing stellar clinker left behind continues to cool and dim, and within a hundred thousand years—peanuts to a star—is completely abandoned by its nebula. Earlier, the star had fused hydrogen to helium, and then, at a much higher temperature produced by gravitational compression, helium into carbon and oxygen. Now the star has shrunk to about the size of Earth and has an average density of a million grams—a metric ton—cm^{-3}. At this density, the electrons in the gas become "degenerate" as a result of the Pauli exclusion principle, which states that no two electrons can have the same properties, in this context, spin, location, and momentum (mass times velocity). Location and momentum are related through the "uncertainty principle," a canon of quantum mechanics that tells us that we cannot simultaneously know the location and the momentum of a particle such that the uncertainty in one times that in the other equals h, Planck's constant (a photon's energy is h times its frequency). Space and momentum each have three dimensions (a particle can move in any of three directions). Within a six-sided box with combined dimensions of position and momentum of volume h^3, we can fit only two electrons, and their spins must be opposite. You can add electrons, but only at ever higher momenta. The effect provides an outward pressure that keeps the star from contracting any further and it stabilizes as a degenerate white dwarf. Only by adding mass can you squeeze the star further. The more massive white dwarfs are the more heavily compressed, and thus are smaller and less luminous—just the opposite of the case with main sequence stars.

White dwarfs are everywhere around us, none bright enough to be visible to the naked eye. Most white dwarfs have masses of around 0.6 that of the Sun, though a few, like the famous companion to Sirius, approach a solar mass. As the mass of a white dwarf increases, so do its internal temperature and the speeds of its degenerate electrons, and above a solar mass the speeds begin to approach that of light. Over 50 years ago, Subrahmanyan Chandrasekhar applied the necessary relativity theory to white dwarfs and showed that the pressure produced by degenerate electrons fails above 1.4 solar masses, leading to disastrous collapse. No heavier white dwarf can exist.

It takes about 15 billion years to kill off a star with a mass of about 0.8 solar. Since the Galaxy is some 15 billion years old, no star of less than that mass has ever had time to evolve. Stars at the lower limit grow nuclear-burning cores—and white dwarfs—of about half a solar mass, the other 0.3 solar mass launched into space. As the mass of the original star increases, so does the resulting white dwarf.

The Sun should make a white dwarf of about 0.6 solar mass, losing 0.4 solar mass in the process. At an initial mass of around 9 solar masses, nuclear burning in the star's core changes the composition to a mixture of oxygen and neon, and at about 11 solar masses the core hits Chandrasekhar's limit. Such a star has lost nearly 90 percent of itself. Since there are far more lower-mass stars than higher, every star in the course of its life ejects on the average about half a solar mass back to interstellar space, mass that is commonly enriched in some heavier elements by its passage through the interior furnaces. There is a cost, however. Out of the recycling process drip the stellar remains, the mass that is forever locked within the white dwarf. The time it takes to cool a white dwarf to under 3000 K is greater than the lifetime of the Galaxy; therefore all those ever made are still there for our appraisal.

At last we start to close the loop. The fresh carbon, helium, nitrogen, and other elements leak out into the Galaxy, helping to enrich it with the astronomer's "metals." Since this new stuff blends with the interstellar gases, the younger the star, the more heavy elements it should have—and in the very broad picture, that is what we

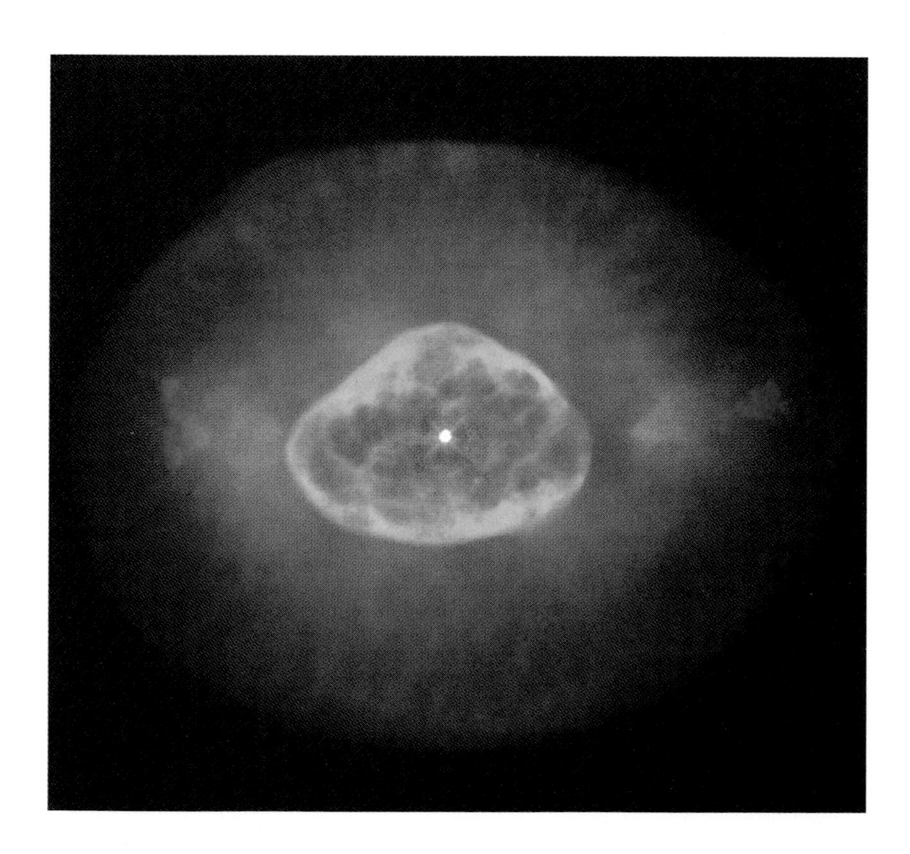

A Hubble view of the planetary nebula NGC 6826 shows wonderful detail that includes currently incomprehensible low-excitation jets (red) running along the nebula's major axis, presumably from matter ejected along the central star's polar axis. The same phenomenon was seen earlier in NGC 6543 (page 184), NGC 7009 (page 187), and NGC 3242 (page 191).

see. Our Galaxy's halo lacks higher-luminosity, shorter-lived stars and must therefore be very old. Relative to those of the disk, the halo's stars are also greatly deficient in heavy elements and in extreme cases contain much less than 1 percent of the metallic content of the Sun. We now see where at least some of the heavy matter that composes the Earth came from. In support of this idea, tiny unprocessed interstellar grains trapped in meteorites have isotope ratios that could have been created only by nuclear processes found in the depths of AGB stars.

The loop tightens. Through the planetary nebulae, the evolving AGB stars produce most of the Galaxy's dust, silicates from the oxygen stars and carbon from the carbon stars, which flees into the cold depths of space. There the grains grow fat on metals and other elements found in the interstellar gas and on volatile ices. They ultimately gather by compressive processes generated by the Galaxy's spiral arms and by the stars themselves to produce the dark interstellar clouds, from which emerge new stars, stars creating yet more stars and providing some of the matter to do it with!

Some of the dust may even come from the destruction of the dying stars' families of comets and planets, presenting us with the possibility of the ultimate in recycling. Could the knots seen in the innards of the Helix Nebula be the result of the dissolution of icy, dusty bodies in the star's Oort cloud? If so, they must have masses much larger than those of our comets, masses comparable to that of the Earth, to produce the observed structures. But given the variations we are now seeing in planetary systems, perhaps that is possible. Could we also, when we walk a smooth patch of Earth, be treading ground that once grew another planet's flowers?

We now return full circle to Herschel's discovery. To avoid confusion, generations of students have been told quite explicitly that the name "planetary nebula" has nothing whatever to do with planets. Yet in one of the wonderful ironies of nature and language, the dust that formed us came in part by way of Herschel's planetaries, a star's Jovian-type planets may have helped shape the nebulae, and its terrestrial planets and even its comets may have contributed to the dust. The planetaries therefore play powerful and varied roles in the begetting of the future planets of some star yet to be.

The recycling process, however, is not done. Though the seeds of the dust from which the stars and planets eventually grow is created by the deaths of these intermediate-mass solar-type stars, they make none of the iron and nickel that constitute planetary cores and river-vaulting bridges, the gold of a ring, the titanium of an aircraft. Things are still missing. We find them as we enter the mysterious world beyond Chandrasekhar's limit.

9

PATHS TO GLORY AND CREATION

≺ *Within the neighboring Large Magellanic Cloud is a nest of bubbles associated with O and B stars. Some bubbles are probably formed by winds, others by supernova shock waves. Still others may be shells of interstellar matter first compressed by supernovae, then ionized by enclosed O stars.*

Riding our theories upward along the main sequence from stars like the Sun, those that produce the white dwarfs, we cross the great divide to the brilliant O stars that illuminate the diffuse nebulae. These stars are genuinely different. Rare, distant, enmeshed in thick clouds, their development is breathtakingly fast and difficult to observe. Their finales are equally distinct; instead of producing the relatively quiet last puffs of planetary nebulae, these stars die through violent explosions. Yet even our own quiet star, which epitomizes the majority of stars, and our own solar world cannot be separated from them: the O stars provide both a driving force for star formation and nearly all the matter out of which the Earth is formed.

MAGNIFICENT PRELUDES

From the mid-northern hemisphere, Sirius shines brilliantly over the southern horizon in crisp winter skies. Drive south to 30° latitude, and up pops the second brightest star in the sky, Canopus; by coincidence the two luminaries lie practically one atop the other. Now travel farther south and also back in time, to the middle of the nineteenth century. At 20°N latitude, you would have seen the *third* brightest star in the sky, and all three lying along a graceful 55°-long curve. For most of the eighteenth and early nineteenth centuries, this star, Eta Carinae, varied between second and fourth magnitude. By 1840 it had begun to brighten, and by 1848 it shone at magnitude –1. It then began a long, slow fade, by 1880 sliding just under naked-eye visibility. And there it has sat for the past century.

Eta Carinae is a superb representative of the rarest of stars, the "luminous blue variables," or, less romantically, "LBVs." Rare they may be, but they are so remarkably brilliant that they can be seen over most of our Galaxy and can easily be identified in others. Relatively nearby ones include P Cygni (for which the famous P Cygni spectrum lines are named) and AG Carinae, found coincidentally in the same constellation—Carina, the Keel—as Eta Carinae. P Cygni is reported to have experienced in the 1600s the same kind of brightening as would Eta Carinae some 200 years later. P Cygni has an absolute visual magnitude of –9, Eta Carinae a maximum of –10, and AG Carinae an astonishing –11, six times brighter than P Cygni (if AG Carinae were 10 pc away it would appear about as bright as the quarter Moon). If invisible ultraviolet radiation is taken into account, Eta Carinae is *3 million times* more luminous than the Sun! The luminous blue variables have an astonishing pedigree, and they and their kind produce equally astonishing progeny, without which none of us could exist.

The Hubble Space Telescope can resolve mass loss from luminous blue variables. The dramatic bipolar flow from Eta Carinae is produced by a 100-solar-mass blue supergiant at its center.

We are now roaming through the main sequence in the ethereal realm between 11 and 120 (or so) solar masses, where stars can no longer make white dwarfs. Analyses of the gravitational effects of double stars show that an 11-solar mass star has a spectral class of B0, slightly cooler than the O stars, but for convenience we will lump all the high-mass stars together under the "O-star" rubric. The evolution of O stars is extraordinarily fast. Energy-generating nuclear fusion is dominated not by the proton–proton chain (as it is in the Sun) but by the carbon cycle, in which carbon atoms grab protons and work their way up through nitrogen and oxygen, finally dropping back to carbon again after absorbing four protons and spitting out helium nuclei. Though these stars have much more hydrogen fuel in their huge nuclear-burning cores, they are so hot inside, and the carbon cycle so efficient, that their fuel is burned at rates so enormous that they live much shorter lives than do lower-mass stars. Theta-1 Orionis C, the great 20-solar mass class O6 illuminator of the Orion Nebula, will take less than 10 million years from its birth to convert its core hydrogen to helium. As it does, it will slowly slide to the right on the HR diagram, leaving the main sequence at about the time it cools enough at the surface to stop ionizing its nebula. Once the central hydrogen disappears, the contracting core will send the star zipping across, at roughly constant luminosity, to the cooler portions of the HR diagram, where such stars normally fire up their core helium. As it does it will expand, becoming first a hot blue supergiant star and then a beautiful *red* supergiant, its size approximating the orbit of Jupiter, and it will dominate a weakly glowing nebula now ionized only by Theta-1's lower-mass companions. Giants are the offspring of stars on the intermediate section of the HR diagram, those between 0.8 and 11 solar masses; supergiants are the brood of the high-mass stars. The diameters of the greatest of them can approach the size of the orbit of Saturn.

Everything in an evolving high-mass star is of majestic proportions. The high luminosities and immense sizes of these stars drive powerful mass loss that ultimately dwarfs anything that the more numerous giants can ever produce. On the far upper main sequence, hotter than spectral class O6 or so (that is, hotter than 40,000 K), the winds continue right from the time of formation, never really stopping, as witnessed by powerful P Cygni lines in the stars' ultraviolet spectra. The winds increase in power while the stars evolve as supergiants and send shock waves into the surrounding interstellar medium, blowing expanding bubbles several parsecs across. Farther out, ionization bubbles—the Strömgren spheres—can eat into the surrounding interstellar medium, where they set up expanding shock fronts. At the largest scale, the combined action of entire OB associations (which include stellar explosions) blows enormous superbubbles that extend for over 100 parsecs.

The highest mass stars, those above about 40 solar masses, are so enormously windy that they, like the Miras, can become completely enmeshed in their own ejecta. Moreover, mass loss is not at all continuous, but extremely erratic. As a result, high-mass supergiants—distinguished as "hypergiants"—periodically bury and then reveal themselves. Rho Cassiopeiae, normally a yellowish fourth-magnitude G hypergiant, faded to sixth magnitude in 1945, its spectral class dropping to reddish M; it took two years to recover. The star had loosed a massive cloud of dusty gas and, until the dust dissipated, buried itself.

Stars of about 50 solar masses and up lose so much matter so quickly that their evolution becomes stalled, and they remain blue hypergiants. These brilliant stars float at the edge of the classical and—for stars—dangerous "Eddington limit." Although photons are massless, they carry energy and momentum. A particle either struck by or absorbing a photon will therefore move or recoil. The constant barrage of photons passing through the star's layers consequently produces an outward radiation pressure. A star automatically adjusts to a size in which in any layer the outward push of gas pressure plus radiation pressure equals the force with which the layer is pulled inward by gravity. Radiation pressure is insignificant in the Sun, but it becomes important as stellar luminosity climbs toward the top of the HR diagram. In the realm of the hypergiants, the outward radiation pressure on the gas dominates, and at the Eddington limit it is so strong that it exceeds the inward gravitational pull at the stellar surface. The star now has effectively zero surface gravity, and mass can easily depart. Such stars will simply tear themselves apart, producing winds that can exceed a rate of 10^{-4} solar mass per year and becoming the luminous blue variables.

Would that a spectrum of Eta Carinae had been taken in the 1840s, when it was at its peak! One taken in 1893 has absorption lines typical of a yellow-white F supergiant. Today in the optical spectrum the star shows us nothing but the emission lines of a surrounding nebula, which is revealed by a *Hubble Space Telescope* image as an enormous tilted bipolar flow. Dust condensing from the outflowing gas has completely blocked an optical view of the star. Infrared observations uncover a small group of stars at the heart of the flow, one star far brighter than the others. Radiation from the surrounding nebula shows that the leader of the pack must be very hot, with a temperature of 25,000 K. A blue supergiant of maybe 100 solar masses, it may be the most massive star in the Galaxy. A century ago, a spectroscopist caught Eta Carinae in the process of being buried; its F-type spectrum was formed not by the star but by absorption in the outflowing cloud. At the time of peak mass loss Eta Carinae may have lost as much as 0.1 solar mass a year and during its outburst a total of one or more solar masses, matter now seen as the surrounding cloud. The mass outflows are highly chaotic, AG Carinae throwing off matter in huge blobs.

AG Carinae, hidden behind the disk, drives mass outward in chaotic blobs.

Ultimately, whatever the exact course of evolution (and we are mightily uncertain of what comes before what), the collection of highest-mass stars, those in the hypergiant range and somewhat below it, can strip themselves (almost?) completely of their hydrogen envelopes to reveal the result of millions of years of nuclear burning. These odd products of stellar evolution, known as Wolf–Rayet stars after Charles Wolf and Georges Rayet, the two nineteenth-century French astronomers who first identified them, have spectra loaded with powerful emission lines. Hydrogen is absent, the stars dominated instead by helium. They are found in two distinct

NGC 6888, a lovely bright ring nebula in Cygnus 6 pc across that was ejected by a nitrogen-rich Wolf-Rayet star, is sweeping up surrounding matter.

flavors, nitrogen-rich (WN) and carbon-rich (WC). The two varieties may represent successive stages of evolution: the N-rich variety, the result of hydrogen burning through the carbon cycle; the C-rich, the result of helium burning.

Whatever the evolutionary scenario, Wolf–Rayet stars are very windy. Many are in binary systems with massive but normal class O companions. Orbital analyses show that the Wolf–Rayet stars have masses of typically 20 solar masses, well below the hypergiant limit. Their companions, however, are *more* massive. For the Wolf–Rayet stars to have evolved first, their original masses must have been larger than those of their companions. The Wolf–Rayet stars have cut their masses at least in half, launching tens of solar masses into space. These massive gems are commonly surrounded by lovely shells of gas known as ring nebulae, which are high-mass kin to planetary nebulae in that they are produced by winds that have ejected the stellar envelopes, sending them crashing into their surroundings. There are, however, major differences (do not, for example, confuse them with the Ring Nebula in Lyra, a real planetary in the shape of a ring; logical nomenclature is not an astronomical strong point). Ring nebulae surrounding O stars are more massive than planetaries, typically containing well over a solar mass of dusty gas. Many planetaries are modestly enriched with the by-products of nuclear burning; *all* ring nebulae, however, are *highly* enriched. We see C-rich ring nebulae from WC stars, N-rich clouds from WN stars, both kinds prosperous with newly minted helium. Planetaries also remain "pristine" for most of their visible lives, amalgamation of their gases with the interstellar medium, or with previous winds from the parent stars, taking place only as they end their short lives. Like all processes in O stars, however, this blending process begins early. By the time the ring nebulae are fully developed, their chemically enriched compositions have already been diluted by the violent mixing of the nebulae with the winds that initially cut the parent stars down to Wolf–Rayet proportions. Returning their matter to the interstellar medium from which they came, these high-mass stars make a generous gift of heavy atoms to be used by the next stellar generation.

COLLAPSE

Physical systems seek their lowest energy states, and under the relentless force of gravity, stars attempt to make themselves as small as possible. This action is the driving principle behind all stellar evolution. Yes, main sequence dwarfs swell to giants, supergiants, even hypergiants, but these expansive states are temporary and superficial phenomena. At the heart of the star, the part that at the end will be all that remains, is a shrinking nuclear-burning core that in the case of white dwarfs is finally stabilized by the outward pressure of electrons.

Nuclear burning is governed by essentially the same principle. Thermonuclear energy is released as a result of the binding together of protons and neutrons. At each stage of burning, the new and improved nucleus is yet more tightly bound, carbon from helium burning tied more tightly than helium from hydrogen burning. The most tightly bound of all atomic nuclei is iron, and that is where nuclear burning is headed. It is prevented from getting there at intermediate stellar masses (where fusion stops at carbon and oxygen) because mass loss and stabilization by electrons keep the temperature from rising high enough. But in the supergiants, where core masses are already near Chandrasekhar's limit, the internal temperature climbs

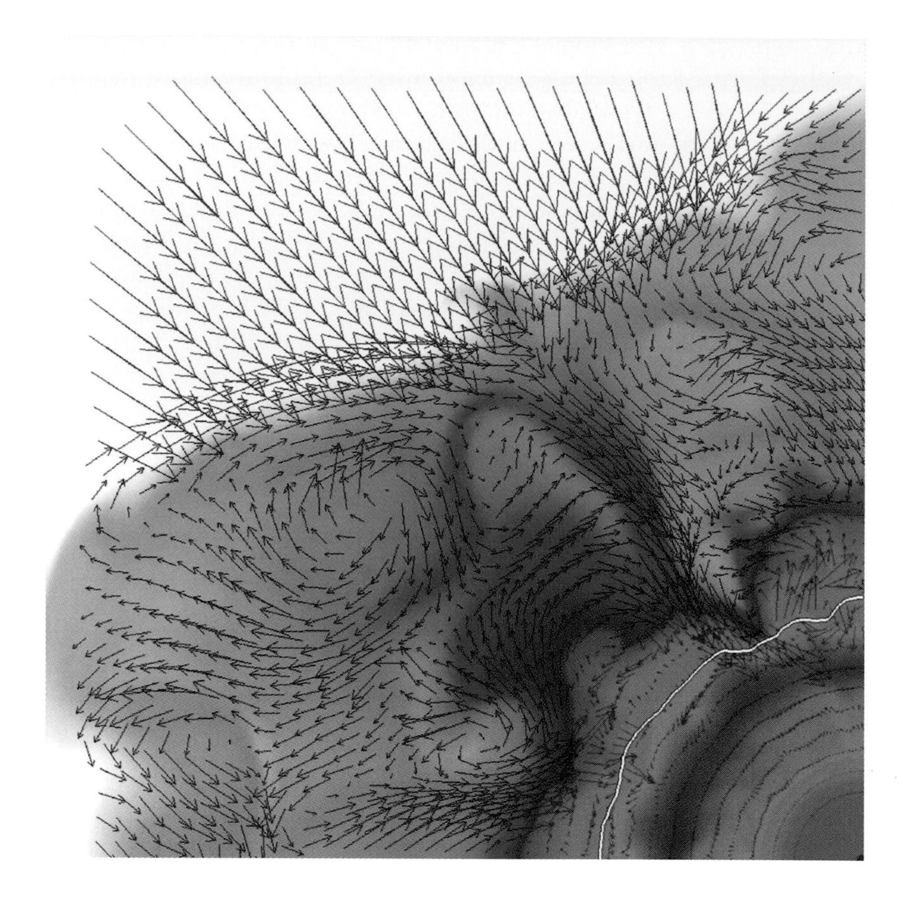

Computer simulation of a supernova 0.089 second after the collapse. Neutrons are forming at the center. The white line, 20 km out, is the "neutrinosphere," the surface at which the inner gas becomes transparent to the neutrinos that help drive an outbound shock. Arrows show the turbulent gas motions that follow the collapse. The surface of the supergiant has yet to learn of its death.

so high that the core does not become degenerate and nuclear burning happily carries on. Main sequence life is terminated when hydrogen is exhausted in the creation of helium. The helium core then contracts within an H-burning shell, as in an ordinary developing giant. When the temperature is high enough, helium burns to carbon and oxygen. But now we reach the break point. The developed carbon–oxygen core can shrink and fire up to burn to a mixture of magnesium, neon, and oxygen, which ultimately shrinks to fuse to silicon (the means is not at all straightforward); the silicon in turn goes to iron. We wind up with an iron core surrounded by concentric shells that are successively burning chiefly Si, Ne/Mg/O, C, He, and H, all enclosed by some sort of envelope, which in the case of ordinary supergiants is made mostly of hydrogen, or in that of the Wolf–Rayet stars of enriched helium.

Aside from minor reactions that generate little energy, the fusing of the iron core into something heavier would *require* rather than release energy, so once this core, now briefly degenerate, starts its contraction it violently collapses: a Chandrasekhar-mass of iron shrinks from 1000 km across to less than 50 km in less than a second at a third the speed of light. In this fierce furnace, where the temperature climbs beyond 10 billion K, the iron is broken down into its component particles, the protons and surrounding electrons shoved together to create a sea of neutrons. As the growing neutron core collapses it generates within itself a shock wave that attempts to move outward. But with insufficient power, it stalls as inward-raining matter keeps the shock from expanding.

Then a remarkable thing happens. The creation of the neutrons produces vast numbers of neutrinos, massless (or nearly massless) particles that are created in many nuclear reactions, including those of hydrogen burning. Neutrinos are the near-ultimate in noninteractive matter. Some 70 billion from the Sun pass through each square centimeter of you—and of the entire Earth—per second. On the average it would take a shield of lead half a light-year thick to stop one. Yet the densities within the star surrounding the collapsed core are so high, some 10^{14} g/cm^3, that the gas is *thick* to neutrinos. They cannot get out, and they push at the outer edge of the core, giving the shock a mighty shove that releases it and allows it to roar outward. As the density drops, the neutrinos escape the star at nearly the speed of light, and a few hours later, when the shock hits the stellar surface, the star rips apart at a speed of about 10,000 km/s — and someone, somewhere, will eventually see a supernova blossom into the night sky. Even though 99 percent of a supernova's energy is carried away by the neutrinos, a core-collapse supernova can reach absolute visual magnitude –18, over a billion times brighter than the Sun and not that much dimmer than its entire host galaxy; it is the single most violent event we know of in nature.

In the cauldron of the explosion, at a temperature now of almost 200 billion K, nuclear reactions go berserk, making a good fraction of a solar mass of radioactive nickel (^{56}Ni) as well as much smaller quantities of other elements all the way through uranium to plutonium and quite likely beyond. The nickel decays through cobalt (^{56}Co) to stable iron (^{56}Fe), the radioactive decay producing gamma rays and eventually much of the exploding star's visible light. What began as nearly a solar mass of core iron thus ultimately generates a tenth or so of a solar mass of the stuff in the envelope that gets blasted into space, aiding in the continuous enrichment of the interstellar medium and someday to be incorporated into new stars. Do you want to touch the debris of a supernova? Pick up a dinner fork.

Not all of the devastated star gets blasted into space. The mass inside the initial shock wave is so dense, its gravity so terrible, that

Supernova 1987A resulted from the collapse of the iron core of a blue supergiant no longer in existence. At its peak the supernova hit third apparent magnitude and, though 52,000 pc away, was easily visible to the naked eye.

a compact ball of neutrons only 20 or 30 km across remains. This "neutron star" contains about 1.5 solar masses or so left behind after the exploded debris dissipates into space. White dwarfs are supported by degenerate electrons that have filled their quantum mechanical spaces. A neutron star is the next stage up, a compact body over a million times denser than a white dwarf that is supported not by electrons but by degenerate neutrons. A remembrance of the initial great O star-become-supergiant, it is a celestial memorial that will endure forever.

As white dwarfs have an upper mass limit, so do neutron stars, the value calculated to be about 3 solar masses. In extreme cases, the collapsed core might exceed the limit, resulting in the fabled black hole. There is nothing to support it, and it collapses forever, its apparent "surface" a tiny sphere only a few kilometers across at which the escape velocity equals the speed of light and from which nothing, including radiation, can escape.

The last supernova witnessed in our own Galaxy occurred in 1604, before the invention of the telescope. What we know about supernova detonations has come largely from observing them in other, usually distant, galaxies. Iron-core-collapse supernovae are the finales to massive stars, and they are consequently observed in the disks of galaxies, where the massive O stars are found. Some have strong hydrogen emission lines that come from the exploding envelopes of ordinary blue and red supergiants. Stars like Betelgeuse, Antares, and Rigel are fine candidates for the next nearby event, though the prime aspirants are the more distant monster luminous blue variables. Eta Carinae could go anytime.

An equal fraction of supernovae, however, have no hydrogen lines in their spectra at all. Some of these are found in galactic disks, and some fraction of these are probably derivatives of the helium-rich Wolf–Rayet stars. However, we also observe hydrogen-deficient supernovae in the outer parts of galaxies, in ancient galactic halos where the high-mass stars have long since died. All that remains there now are low-mass stars that cannot possibly develop collapsing cores.

This second path to explosive glory is still not cleared of underbrush, and our view is uncertain. There are a variety of possibilities for this kind of hydrogenless supernova, all involving double stars and white dwarfs. Of two main sequence stars in a binary, the more massive must evolve first. If the members of the pair are sufficiently close, the growing giant may expand past the less massive component and wrap it in its tenuous outer atmosphere, so that, for a while, one star actually orbits inside the other. Friction then slowly brings the stars closer together as the giant develops a planetary nebula and shrinks into a white dwarf. At the end, a white dwarf and a main sequence star are in close proximity, perhaps close enough for tidal distortion of the main sequence star to cause matter to flow onto the smaller star. When enough fresh hydrogen has accumulated on the white dwarf, its surface explodes in a thermonuclear bomb that drives the star to great luminosity, right to the Eddington limit, and a "nova"—a "new star"—appears in the sky. These events are quite common in the Galaxy, their brilliance reaching absolute visual magnitude –8 or –10, comparable to the brightest O dwarfs and M supergiants. The explosions do little harm to either of the stars involved, and the debris, enriched with yet more elements created in the fusion blast—helium, nitrogen, carbon, neon, magnesium, silicon—are sent into space.

However, what if the white dwarf comes from a massive progenitor and is near the Chandrasekhar limit? Then the stolen accumulated matter might push it right over the edge, and a runaway thermonuclear reaction in the carbon-rich core results. This kind of supernova brightens even more than the core-collapse variety and may destroy itself altogether, in the process launching yet *another* load of iron—perhaps a few tenths of a solar mass or so—and other elements into the interstellar medium. Such a scene should not happen often, and supernovae are indeed rare.

Now take evolution process one step further. Say the white dwarf is well under the Chandrasekhar limit. The ordinary dwarf now expands as a giant, enclosing the white dwarf, and the two are drawn yet closer. General relativity theory predicts that an accelerated mass should generate gravity waves. As the tiny stars orbit, energy released through the radiation of such waves causes the pair to spiral even closer until they merge, the product of the two again exploding in a thermonuclear runaway.

Even an ordinary white dwarf well below the Chandrasekhar limit might be induced to go off if it can accrete enough matter from a binary companion, one that had previously been stripped to its helium-rich interior by tides raised by the white dwarf's progenitor. After a couple tenths of a solar mass have been deposited on the white dwarf, the fresh helium can ignite and send a pressure wave inward, again kindling the white dwarf's core and destroying the star in iron-bound explosion.

"Extraordinary Events . . . Extraordinary Evidence"

What led to these remarkable pictures of natural violence? In response to assertions that unidentified flying objects are the craft of "space aliens," the UFO debunker Philip Klass suggested that "extraordinary events require extraordinary evidence." Extraordinary evidence (indeed, any solid evidence) is lacking for extraterrestrial space ships, but not for exploding stars. Observational evidence dates to 1572 (just 30 years after Copernicus had forever displaced us from the center of the Universe), when Tycho Brahe gazed into the Milky Way in Cassiopeia and carefully described a dazzling "new star" that eventually took his name. Tycho's Star, at its peak visible in daylight, rivaled Venus in apparent brilliance and was seen for over two years. Just 32 years later, in 1604, a similar event in the constellation Ophiuchus was studied by Johannes Kepler, the discoverer of the laws that govern planetary motion. Then in 1885 a luminous pinpoint erupted in the heart of the Andromeda Nebula (M 31), almost reaching naked-eye brightness. Once Edwin Hubble determined M 31's distance in 1924, astronomers realized the "new star" seen 40 years before must have been enormously bright, vastly more so than a common nova, in a class with Kepler's and Tycho's stars (which even in apparent brightness had also far outstripped the novae). For these extraordinary events, and others seen in more distant galaxies, Fritz Zwicky and Walter Baade of Mount Wilson

Observatory coined the term "supernova" in 1934. In 1936, Zwicky, in one of the most outstanding astrophysical deductions of all time, suggested that they were caused by the "rapid transitions of ordinary stars into neutron stars."

The older historical record is equally clear. The Chinese and Japanese of the past two millennia kept meticulous records of celestial events and of new, seemingly temporary, stars. Many of these phenomena were obviously comets, while others behaved as ordinary novae. But a whole class stands out as unusually brilliant, comparable to Tycho's and Kepler's stars; with good probability, these include "guest stars" of the years 185, 386, 393, 1006, 1054, and 1181. The visitor of 1006, seen in our constellation Lupus, was the brightest stellar object known in all history, reaching apparent magnitude between –8 and –10, comparable to the brilliance of the quarter Moon. Though from China it was never far above the southern horizon, it was clearly visible in the daylight sky and at night cast strong shadows: a contemporary text states, "It shone so brightly that objects could be seen by its light." The event of 185 was not far behind.

The supernova of 1054 must have impressed people everywhere. Here it is apparently drawn on a rock, alongside the crescent Moon, in Chaco Canyon, New Mexico.

Recent times have yielded more technical testimony to stellar catastrophe. A century after the M 31 supernova of 1881, an image of the region taken with a filter that passed only light of an iron absorption line showed a black spot created by the outblasted iron. And although no supernova has been seen in our Galaxy since 1604 (we are certainly about due), one was close: in 1987 a blue supergiant erupted in the Large Magellanic Cloud. Detectors in Japan and the United States captured the predicted number of neutrinos from the core of Supernova 1987A a few hours before anyone saw the star erupt; our instruments had looked directly into the middle of a collapsing star. Astronomers also observed gamma rays from the decay of radioactive nickel into iron. We have our extraordinary evidence.

The Wreckage

In 1731, the English astronomer John Bevis, creator of the star atlas *Uranographia Britannica,* discovered a fuzzy cloud just off one of Taurus's horns; Charles Messier, referring to the "English Atlas," listed the cloud as number 1 in his famous catalogue. A visual examination by William Parsons, Lord Rosse, in the 1840s, prompted him to call it the "Crab Nebula." Photographs from our own era show crablike legs entangled in a smoother amorphous background. As early as 1921 the positional coincidence between the Crab and the Chinese guest star of 1054 suggested that the Crab was the debris of the brilliant interloper, a relation confirmed in the 1940s by observation of the Crab's expansion. Over a period of years we can watch the filaments in the Crab Nebula move outward. Working backward at the current rate, astronomers calculate that all the filaments should have begun expanding from a point near the center about the year 1140, a good match with the historical record considering the inevitable observational errors and possible accelerations.

The amorphous background of the Crab radiates a continuous spectrum, while the filaments produce a rich emission spectrum with the usual recombination lines of hydrogen and helium and forbidden lines of oxygen, nitrogen, sulfur, neon, and other elements. A combination of spectral and positional measures allows us to know the Crab's distance. As they do for planetary nebulae, Doppler shifts in the line spectra yield the Crab's expansion veloc-

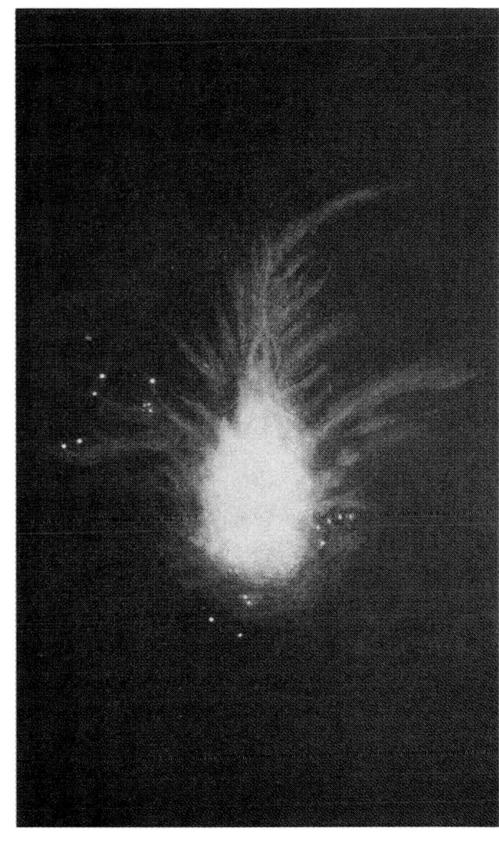

The Crab Nebula (left) is the 4-pc-wide exploded debris of the supernova of the year 1054. At right is an early drawing of the Crab Nebula made by Lord Rosse.

ity, which is 1500 km/s. Assume the nebula is expanding with the same velocity *across* the line of sight. The angular motions of the filaments depend on both distance and velocity; with velocity known, we can easily calculate the distance. From more realistic assumptions about the nebula's three-dimensional shape and the expansion's asymmetry (based on the directions of the local magnetic field and the rotation of the remnant star), a distance of about 2000 pc and a diameter of 4 pc is derived. The star must have reached absolute visual magnitude –16, radiating the light of a quarter of a billion suns—clear confirmation of a supernova event. The hydrogen emission lines establish that what the Chinese astronomers had watched was the collapse of an iron stellar core. The Crab, the remnant of the grand explosion, is at least one form of "supernova remnant," or "SNR." There must be many others.

In the early days of their science, radio astronomers began listing newly discovered sources of celestial radio radiation—"radio stars"—by constellation. So many were found so quickly that the naming system almost immediately broke down, to be replaced by catalogue numbers; nevertheless, the names Virgo A, Sagittarius A, Sagittarius B, and several others that identify the brightest objects endure today. At the time, the identification of the radio sources' optical counterparts was a tricky business because some sources radiate weakly at short wavelengths and because an ordinary radio telescope's resolving power—its ability to establish precise position—is limited. With improved instrumentation, which included radio interferometers (two or more separated radio telescopes that work as one, allowing great directional ability), astronomers matched Virgo A with the huge galaxy M 87 and Cygnus A with a far more distant galaxy; they realized that Sagittarius A and B are respectively the galaxy's center, which probably contains a massive black hole, and a dense star-forming region (because of interstellar dust, neither source is actually visible optically). In 1949, they nailed down Taurus A as the Crab Nebula, which at some wavelengths is the brightest radio source in the sky.

Diffuse and planetary nebulae all present continuous spectra beneath their emission lines. The optical and ultraviolet continuum is caused principally by the recapture of free electrons into various orbits of hydrogen and by the occasional emission of two photons instead of one (Lyman α) in the electron's jump from orbit 2 to orbit 1. The infrared is often dominated by radiation from warm dust, while the radio spectrum is ruled by free–free emission (weakly present in the optical), wherein free electrons are merely braked rather than captured while passing free protons. All these processes are well understood, allowing astronomers to predict the continuum strength throughout the entire spectrum.

These rules failed miserably for the Crab. At first no one was able to reproduce mathematically the powerful radio emission on the basis of the strength of the optical continuum. In a series of steps, astronomers gradually realized that the continuum was similar to that produced in the kind of atomic accelerators called synchrotrons by radiation from electrons moving near the speed of light and trapped in magnetic fields. The electrons are deflected into spiral paths around the magnetic field lines, the accelerations causing the electrons to radiate in the directions of their motions. "Synchrotron radiation" behaves according to a power law, the strength of radiation inversely proportional to a low power of the frequency. As a result, the strength of synchrotron radiation rises quickly as frequency decreases (and wavelength increases) into the radio domain. Synchrotron radiation immediately identifies supernova remnants, powerful magnetic fields, and some kind of mechanism that can produce fast electrons. It is observed in several other contexts and, notably, is emitted from Jupiter's magnetosphere by electrons caught from the solar wind. About 200 galactic supernova remnants have been identified by their radio emission. Half of these are associated with some kind of optical features, which may be no more than a few faint gaseous shards, and some three-quarters are observed in the X-ray part of the spectrum.

The emission-line spectrum from the Crab's filaments at first appears similar to that from diffuse and planetary nebulae, but it is not quite the same. First, there is no immediately obvious hot ionizing star. The emission-line strengths are also peculiar, in particular those of low ionization like [S II], forbidden lines from singly ionized sulfur. We can replicate the line strengths theoretically by assuming that the filamentary gas is ionized by shock waves and by the powerful synchrotron radiation, which extends from the radio through the optical all the way into the hugely energetic gamma-ray spectrum.

The Crab Nebula must have an awesome power source. Even in the 1930s, Zwicky had suspected that the engine behind the Crab was a neutron star and even identified it as a peculiar star in the center of the object that had no absorption lines; but there was no way to prove it. Demonstration came—serendipitously—from another direction. In 1967, the British astronomer Anthony Hewish set up a specialized antenna near Cambridge to observe point radio sources as they "twinkled" because of variable refraction induced by the solar corona. The monitoring of the experiment was given to a graduate student, Jocelyn Bell, who one morning found odd-looking pulses spaced 1.3 seconds apart on the continuous automated ink recording. The pulses disappeared for two months then suddenly reappeared precisely on the previous schedule, with the same clocklike period of 1.337011 . . . seconds. After some initial speculation about the possibility of having discovered some kind of interstellar communications beacon, the discovery of several more of these radio-emitting pulsars, including one in the Crab Nebula that was flashing 30 times per second, led theoreticians to postulate radiation by rapidly spinning stars. The short periods implied very small diameters of only tens of kilometers, which in turn implied extraordinary densities of 10^{14} g/cm^3, essentially that of nuclear matter.

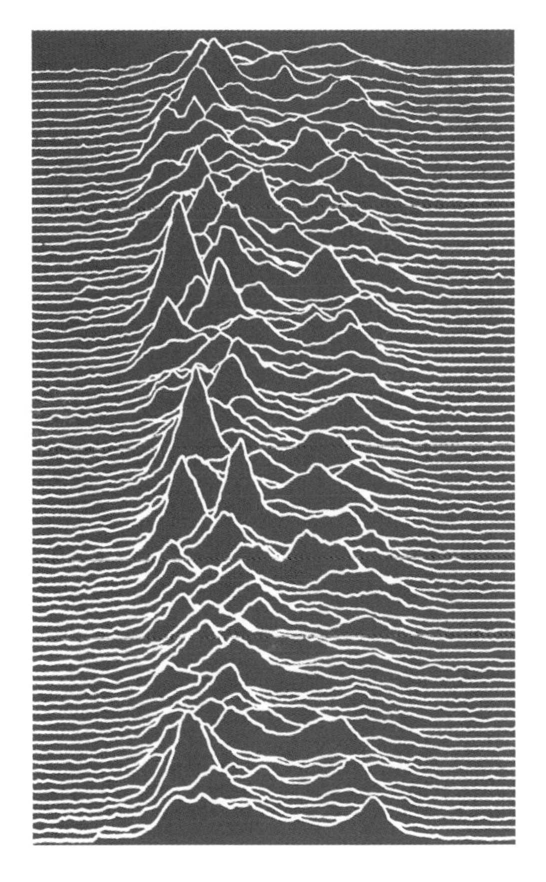

Successive bursts of radio radiation from the first known pulsar (each read from left to right) are stacked from bottom to top according to the known average period of 1.33739 seconds. Each pulse exhibits considerable complexity, but over time the average pulse shape remains the same.

Pulsars must therefore be Zwicky's neutron stars. Two years after Bell's discovery, the Crab pulsar was located optically: it was the star Zwicky had pointed out a generation before. The Crab pulsar was eventually seen radiating across the entire electromagnetic spectrum, even into the gamma-ray domain.

Over 600 pulsars are now known, about a dozen of them identified with supernova remnants. A pulsar is the collapsed remains of a supernova, spun up by simple conservation of angular momentum. The collapse also concentrates the star's magnetic field, which, like most astronomical magnetic fields, is tilted relative to the neutron star's rotation axis. Magnetically accelerated electrons then produce radiation along the field axis, and if the Earth is in the way of the wobbling field, we see a short blast of radiation much in the way we see the rotating beam from a flashing lighthouse. A pulsar is at first very energetic, rotating and pulsing rapidly, throwing out both high- and low-energy radiation. Radiation generated by the magnetic field draws energy from the star, and it slows with time. As it spins down and its energy decreases, high-energy optical pulses are no longer emitted; and the little body, now pulsing with a period of a second or more, appears only in the radio spectrum and eventually disappears from view altogether.

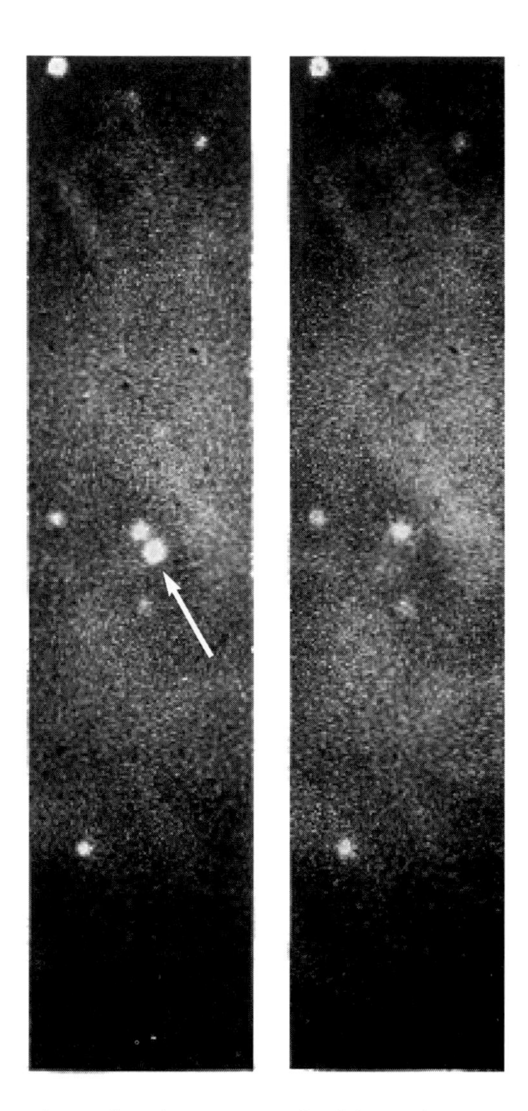

The Crab pulsar appears "on" (arrow) and "off," as it is seen at intervals of 0.015 second. The pulsar completely disappears between pulses, when its radiation does not strike the Earth.

In addition to its pulses, the Crab's neutron star produces a powerful wind of highly accelerated electrons that spread out into the nebula along with the magnetic field to produce its amorphous synchrotron radiation. The inner portion of the expanding nebula is filled with energetic and elegant detail that includes bipolar jets, a shock front formed when one of the jets hits ambient matter, and a circular halo. All these features change and dance around over a period of only months as the engine in the center chaotically varies its energy output. The filaments that make the "crablegs" are the shredded remains of the star's exploded envelope. They are 90 percent helium, as would be expected from a supergiant that had already stripped away part of its envelope through winds. They are also heavy with other elements, though the enrichment of the expected iron, for unaccountable reasons, seems low.

Of the known supernova remnants, the Crab, the remains of a core-collapse supernova, turns out to be unusual, consisting of a filled ellipsoid; almost all the others, which are much older, are hollow shells. Many of them, including Tycho's and Kepler's stars, are known from the way they dimmed to be of the white-dwarf type. In most SNRs we see not so much the stellar leavings of the explosion as the shock wave of the great blast as it propagates through, and

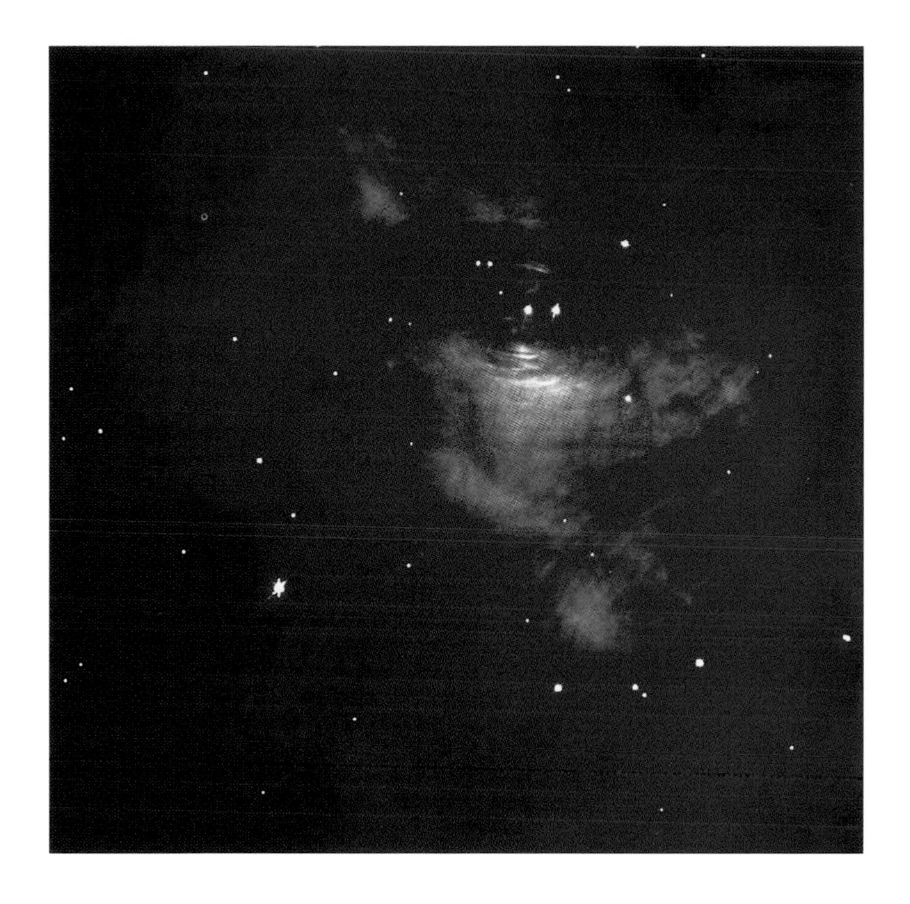

A knot of gas emerging from what appears to be the rotation axis of the Crab pulsar can be seen in this Hubble image. In the other direction, the axis is topped by some sort of cap or halo. X-ray emission, not shown here, comes from a torus wrapped around the axis. The scene changes, flickering madly, over a period of months.

heats, the interstellar medium. Temperatures in the shock wave reach into the hundreds of thousands of kelvins, producing the powerful X rays observed from about half the sources. Eventually the great bubbles expand into the general interstellar medium, where they, along with the bubbles produced by energetic O-star winds, produce the hot coronal gas that surrounds the cooler clouds where star formation takes place.

Most SNRs do not contain pulsars observable from Earth, as the pulsars' magnetic axes are not aligned suitably, and most pulsars are not clearly related to SNRs because they long outlive the expanding clouds of debris, spinning ever more slowly until they disappear from view. In a few instances, however, these old pulsars can be reborn. There is a small subset of pulsars with periods measured not in seconds but in *thousandths* of seconds. Such periods are far too fast to result from an iron-core collapse. Astronomers are almost certain that these "millisecond pulsars" are the result of binary

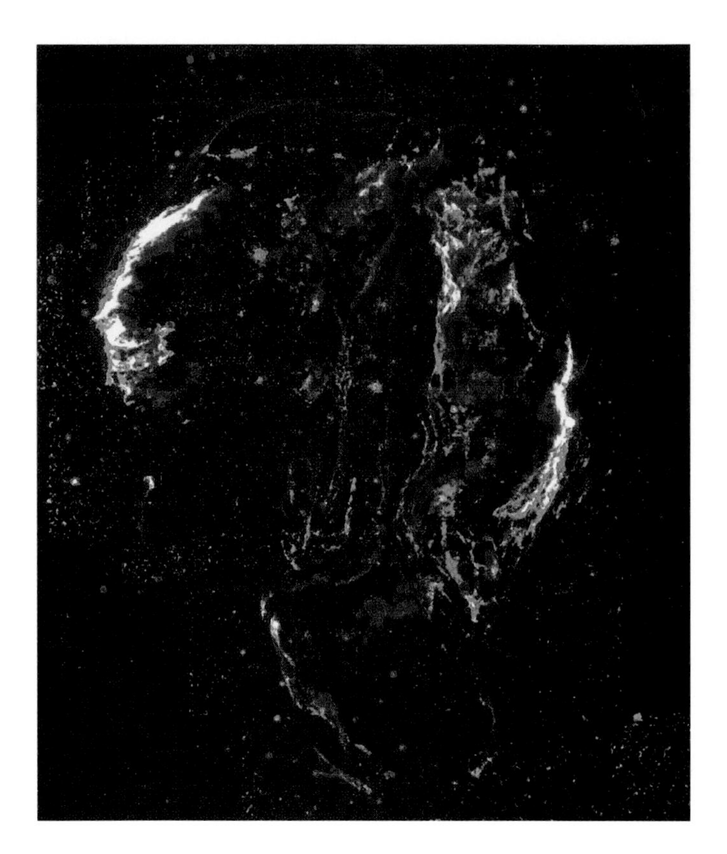

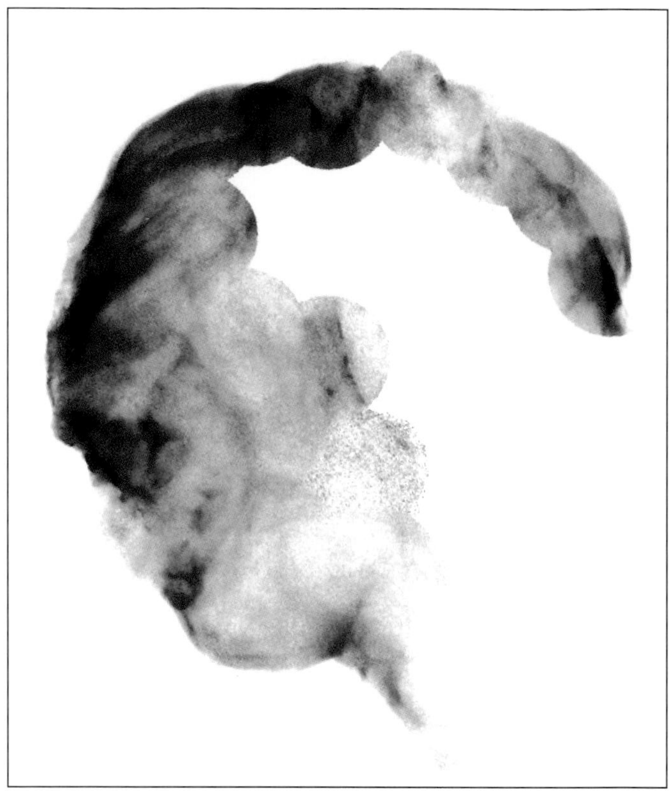

The Cygnus Loop (above), 2° across, is the shock-heated remains of a supernova that detonated some 100,000 years ago. X rays (at right), produced by shock-heated gas, nicely match the optical radiation from the loop's left-hand arc.

action, in which a star transfers mass to a neutron-degenerate companion, one that might once have been a pulsar but has slowed down. As matter rains down on the old pulsar's surface, it also transfers angular momentum, making the small star spin faster than it ever could on its own and giving it a new bloom of life. At the same time, the powerfully radiating neutron star evaporates the unfortunate contributor, eventually leaving us with a single whirling dervish.

Analysis of the pulse periods of a particular millisecond pulsar reveals variations that can be produced only by small orbiting bodies with masses a few times that of Earth that are pulling the pulsar back and forth. This pulsar has *planets,* the first alien planets ever found. Normal planets formed along with the star from the collapsing interstellar clouds should have been destroyed in the course of supergiant evolution or in the explosion. Most likely these planets formed from the debris of the evaporated companion, perhaps the ultimate in celestial regeneration. The implications for extrasolar planetary science are vast: given a source of matter, planets will apparently form anywhere.

Home Once More

Supernovae have three principal and powerful effects that relate directly to us: the compression of the interstellar medium, the generation of cosmic rays, and the production of heavy elements. All these phenomena are crucial to the formation of stars and planets, and all are needed for the creation of the Earth.

Stars cannot form unless a portion of the interstellar medium is so compressed that it becomes gravitationally bound and thus is forced to collapse in on itself. O stars provide an obvious source of compression, from their first powerful winds to their concluding supernova blasts, which (together with white-dwarf supernova detonations) shock the interstellar medium into huge bubbles that strike interstellar clouds and squeeze some of their mass into the dense cores that begin the lives of the stars. In the Large Magellanic Cloud, a site of very active star formation and host to large populations of O stars and SNRs, we see multiple generations as winds and supernova bubbles produce stars that explode to make yet more stars. The shocking force of a supernova is so great that its sudden pressures acting on ordinary giant-star carbon grains create the vast amounts of observed interstellar diamond dust. Diamond dust is found in meteorites, within the debris from which the planets were constructed, firmly tying Earth's formation to the great blast waves that propagate through the interstellar medium.

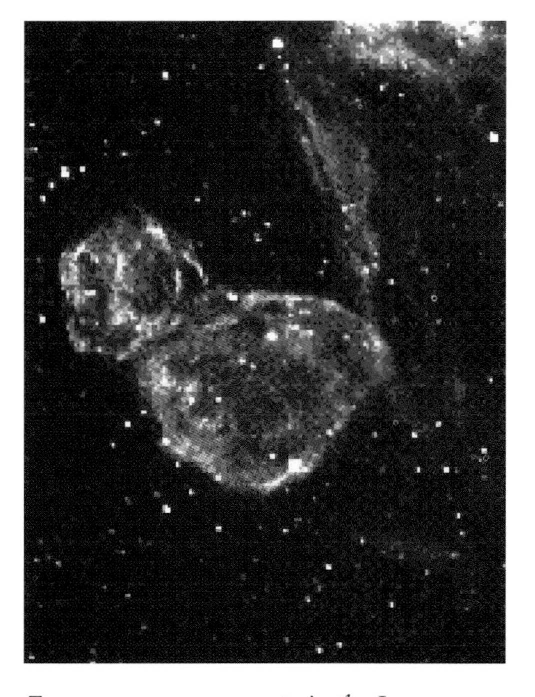

Two supernova remnants in the Large Magellanic Cloud appear to be colliding, demonstrating how interstellar space might be filled with a tapestry of hot tunnels, their compressing shock waves helping to create stars.

Cosmic rays, the high-energy nuclei that travel through space at speeds near that of light, weakly ionize dark clouds to help activate a complex chemistry; the ionization also provides a handle for the Galaxy's magnetic field to grasp, and the result is the removal of some of the clouds' angular momentum. Far from being an obscure phenomenon, cosmic rays collectively contain a third of the Galaxy's energy, comparable with that bound up in electromagnetic radiation and in galactic magnetic fields. Modern satellite-based studies show that the atomic composition of cosmic rays is very much like that of the Sun. In spite of their high energies, they therefore must be made in the same way as ordinary, low-energy matter.

Most cosmic rays are almost certainly the product of supernovae. Within our Galaxy only supernovae (or poorly understood acceleration mechanisms in double-star systems that contain collapsed neutron stars) have enough energy to produce them. They apparently ride supernova blast waves like surfers skimming water waves, and are accelerated to nearly the speed of light by repeated reflections within magnetic fields at the outer edges of the outbound shocks. The Galaxy's magnetic field then traps them, spinning them into huge galactic orbits, and a tiny number crash into Earth to throw down the invisible atomic showers that constantly rain upon us. The high end of the cosmic-ray energy spectrum is thought to come from violent processes that may involve entire distant galaxies.

Finally, although intermediate-mass stars eject considerable amounts of helium, nitrogen, and carbon into interstellar space via planetary nebulae, they are not capable of creating much in the way of heavy elements because the temperatures of the progenitor cores simply cannot become high enough. Except for these three elements, effectively all the heavy stuff in the Universe, including all the iron, has been made in supernovae. Over the past two millennia the historical record lists eight supernovae within our Galaxy. Radio astronomers added Cassiopeia A, calculated to have gone off about the year 1650 but invisible to us because of interstellar absorption. On the average, we see a supernova every two hundred years or so. Like Cassiopeia A, most supernovae are hidden by dust: there is no way we could see even the brightest supernova if it were in the galactic plane on the other side of the Galaxy.

Assume a supernova every fifty years. The Galaxy is about 15 billion years old, so (assuming an admittedly unrealistic constant rate) it should have witnessed a quarter of a billion supernovae. Averaging the different kinds of events and progenitor masses, O-star evolution and supernovae (including the white-dwarf variety) should thus have returned 10 or so solar masses per star back to interstellar space for a total of around 3 billion solar masses, over 1 percent of the stellar mass of the whole Galaxy; a large amount of this matter, an average of several tenths of a solar mass per explosion, has consisted of heavy elements. Supernovae can easily have produced all the iron found in the interstellar medium.

A third the mass of our planet is iron. Most of the remainder is nickel (also in the core), silicon, and oxygen. Our neighboring terrestrial planets, and the cores of the giant planets Jupiter, Saturn, and the rest, have similar compositions. Effectively, all the Earth's mass of 6×10^{24} kg had to have been created in supernovae. If the quarter-billion supernovae of the Galaxy's history had contributed equal amounts (certainly an oversimplification, but a benign one), each supernova gave our planet 2×10^{16} kg of matter, an amount within a factor of 10 equal to the mass of the greatest mountain on Earth, the Big Island of Hawaii. Meteorites also contain the daughter products of the radioactive decay of plutonium. Nothing but a supernova could have had the energy to make it and place it there. Plutonium decays quickly, so the supernova must once have been nearby. Our own Sun may be the result of compression by a specific supernova event that took place some 4.5 billion years ago. The paths to glory and creation end right here.

10

Cosmic Cycles

The Hubble Deep Field shows some 2000 galaxies within an area only 2 minutes of arc across—a fifteenth the angular diameter of the full Moon. Many are interacting or even colliding. They are at vastly different distances and nowhere near as densely packed as implied by this view, a projection on the plane of the sky. The farthest galaxies are many billion light-years away: the Universe was young and these galaxies are full of developing stars.

Where does a circle begin? Interstellar clouds collapse to form stars; the stars then toss dust and enriched matter back into the cosmos, matter that will someday be engulfed by new clouds for the creation of later stellar generations. Where do we actually start? Though in simple form the story is circular, the actual loop of stellar creation is not. It contains multiple twists, turns, interconnections, gaps, and subloops, and it has an entry point through which we can look to the outside.

Opening the Loop

The great cosmic recycling engine, which takes stars from the interstellar gas, tears them apart, and puts their matter back again, is now in place and running fairly smoothly. Like any engine, it had to be started. The ignition key, however, has yet to be found. The Big Bang, the expansion event believed to have created the present Universe, should have delivered into the cosmos nothing but hydrogen, helium, and a little lithium, but none of the heavy stuff of modern stars and planets. Consistently, we see that the older populations of the Galaxy, those of the galactic halo (including the ancient globular clusters), are highly deficient in heavy elements, a few stars having iron abundances as low as a hundred-thousandth that of the Sun. By the time the Sun was born, perhaps 10 billion years after the dawn of the Galaxy, earlier generations had created enough heavy elements to bring the metal content of the Sun beyond lithium up to nearly one atom out of a thousand.

But though we find stars that are very low in metals, we have yet to find stars with *no* metals. We must open the loop and look outward to speculate and theorize. For reasons still inaccessible, the gaseous matter of the expanding Big Bang Universe condensed into self-gravitating blobs that would eventually become clusters of galaxies and then the galaxies themselves. Our Galaxy (as well as all the others) should have begun as nothing but a contracting cloud of hydrogen and helium, as *only* "interstellar matter," though there were as yet no stars to define it as such.

In an astronomical analogue to the responsibility-avoiding "Mistakes were made," we tell our students, "During this collapse the first stars formed." We have little idea how. Without stars, there could be no windy-star or supernova compression triggers. The developing galactic systems were close to one another in the early days, however, and compression could have been provided by the gravitational effects of collisions, mergers, and near misses. But without heavy elements, there could be no dust and no star-forming regions as we know them. Yet these mysterious first-generation stars indeed formed (the proof is our own existence), and it probably would not have taken many of them to have seeded the new Galaxy with enough heavy elements to initiate the star-formation process we see continuing today. As evidence for this hazy picture, the relative abundances in the most metal-poor stars are about those expected from input by supernova blasts.

Once we get started the pathway is clearer. In the simplest picture, as the nascent Galaxy continued to contract, it formed the first metal-poor globular clusters and other halo stars. At the same time, the conservation of angular momentum made the new system spin faster and faster, the rotation slowly flattening it, leaving the globulars behind. As the metal content built, the system flattened further, the next generation of stars and clusters, now distributed in a fat disk, having higher metal contents. As evolution continued, the Galaxy flattened into its present thin, dusty, metal-rich, star-forming disk, surrounded today by successively fatter, older, metal-poor halos.

Though there is some truth to this simple scenario, the complete view is soberingly complex. Except for the extremes, at which the oldest stars of the halo are the most metal-poor and the youthful ones in the disk relatively metal-rich, the correlation between age and metal content is surprisingly weak. The Galaxy seems to have developed spasmodically in a series of bursts of stellar births, the birthrates different in different parts of the system. More important, the Galaxy may in part have developed as a result of mergers with other systems that were in different states of evolution or had even proceeded on different evolutionary pathways. (For example, our nearby companion, the Large Magellanic Cloud, contains young globular clusters, and many galaxies have no disks and little active star-formation activity.) Our Galaxy may in fact be a composite of several galaxies, the lack of good correlations between metals and age direct evidence of collisions long since complete. Even a cursory view of the Universe of galaxies, especially within clusters of galaxies, shows how frequent such collisions are.

Moreover, the basic scenario does not factor in the effects—whatever they may be—of "dark matter." If the majority of the galactic mass were concentrated inside the solar orbit, a star or interstellar cloud orbiting at, say, 16 kpc rather than our 8 kpc would gravitationally feel about the same mass pulling on it and would orbit more slowly in accord with Kepler's (and Newton's) laws. But it does not. The Galaxy's rotation curve stays at about the same velocity as far out as we can observe. A star at 16 kpc must therefore feel twice as much mass as does the Sun, and a star at 24 kpc three times as much. However, only 20 percent of the Galaxy's mass outside the solar circle is in stars and standard forms of interstellar matter. We cannot see what produces the extra gravitational force. Whatever it is, and at this point we have little clue, it is referred to as dark matter. Stars and normal interstellar matter are by contrast referred to as "bright matter," even the dark globules, since although they are dark optically we know what they are and what they are made of. Though not often so stated, "dark" as much connotes "unknown" and "bright" means "revealed."

The galactic disk eventually becomes so thin we can no longer trace it. Does the dark matter extend farther? We can look at the radial velocities of ultradistant globular clusters and nearby dwarf galaxies, members of our own local cluster (which contains some three dozen galaxies within a volume about 2 million pc across) that are gravitationally affected by our Galaxy. We find they are moving too fast for the amount of galactic bright matter we can see. The dark matter seems to extend in a huge, perhaps spherical dark-matter halo that encloses—and may be 10 times the mass of—the visible Galaxy. This halo may extend to a distance of 100 kpc from the center.

We do not know what the dark matter comprises, though candidates abound, including undetected red dwarfs, brown dwarfs, countless Jupiter- and asteroid-sized bodies, uncountable numbers of invisible black holes, huge numbers of neutrinos created in the Big Bang, exotic massive atomic particles as yet undetected in the laboratory as well as in space—or a combination of some or all of these possibilities. Yet in spite of years of looking, in our Galaxy and in others (where we observe the same gravitational effects), no one has yet learned the true nature of dark matter. Equally important, we do not yet know the role dark matter has played, and may play today, in forming galaxies and stars. If nothing else, the existence of dark matter warns us to be cautious about believing too deeply in our theories of creation, theories developed in the face of our ignorance of the nature of 90 percent (maybe, if some calculations of the processes that took place in the Big Bang are correct, even 99 percent) of the mass of the Universe. That said, we continue as if dark

---

Stars are born as a result of a number of interconnected feedback loops chiefly controlled by the Galaxy as a whole. Time runs downward, the flow of events indicated by arrows. (The relative speeds of evolution are not to scale.) Dashed lines show the return of mass to the interstellar medium, wavy lines compressive forces, and dotted lines the routes taken by cosmic rays.

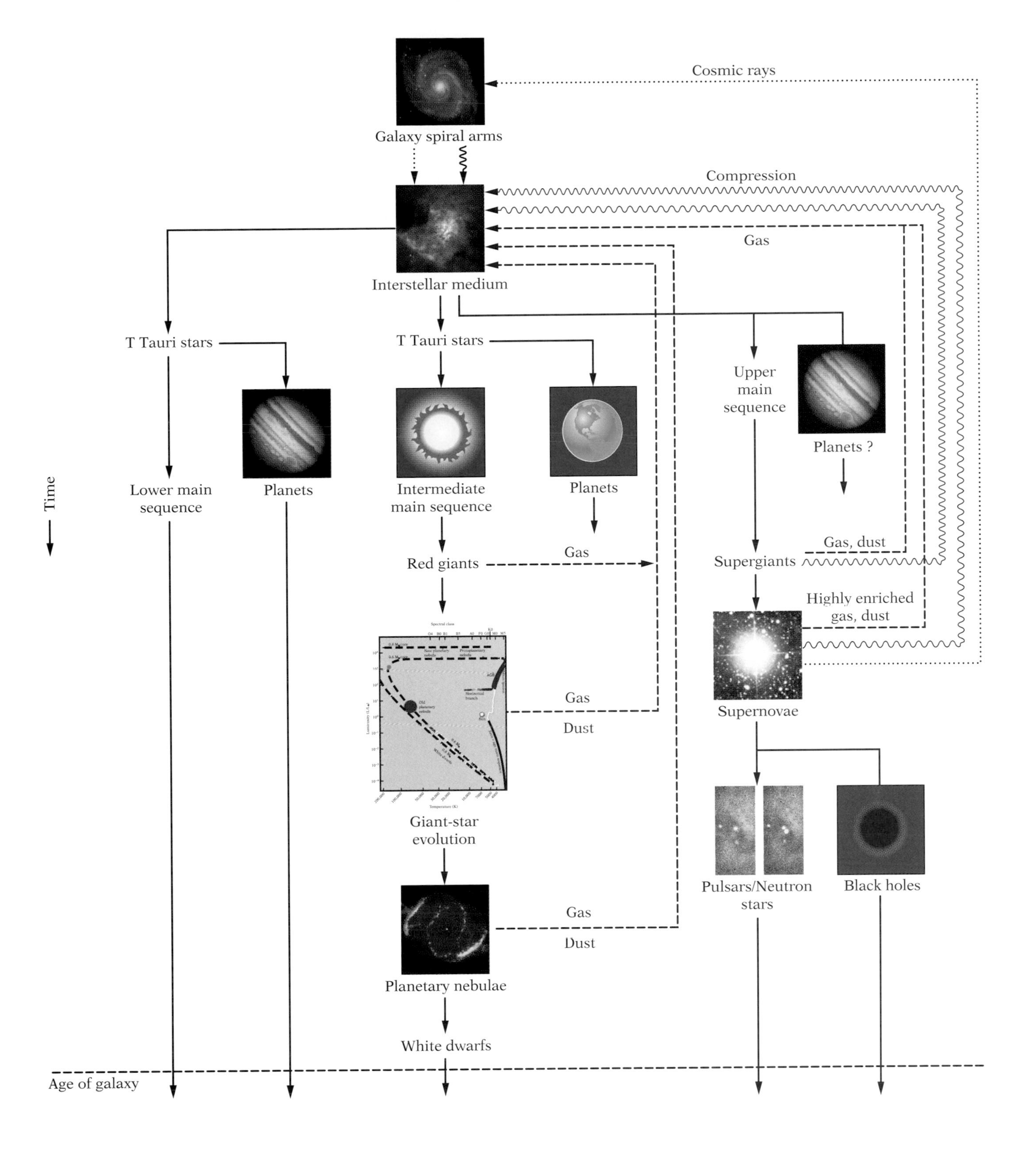
Cosmic rays
Galaxy spiral arms
Compression
Gas
Interstellar medium
T Tauri stars
T Tauri stars
Upper
main
sequence
Planets ?
Time
Lower main
sequence
Planets
Intermediate
main sequence
Planets
Gas
Red giants
Gas, dust
Supergiants
Highly enriched
gas, dust
Gas
Dust
Supernovae
Giant-star
evolution
Pulsars/Neutron
stars
Black holes
Gas
Dust
Planetary nebulae
White dwarfs
Age of galaxy

matter is of little consequence in dealing with star formation and cosmic recycling: that is all we can do.

Synergy

Only within a galactic mass is the density of matter high enough for stars to form. As a result, stars clump together into galaxies, a simple concept but one that should be explicitly stated and appreciated. Though the cycle that connects stellar and interstellar forms of matter is complex, all its parts—including the dark matter—work together, choreographed by gravity in a grand dance.

In a galaxy like ours, the density waves that make the spiral arms seem to be the chief source of the compression of the interstellar medium, that needed to produce the giant molecular clouds. As the waves spread outward, stars form within the compressed gas and dust. Bright O and B stars born within the arms outline the graceful spirals with their bluish color and bejewel them with the reddish diffuse nebulae they illuminate. The O stars produce winds and eventually explode, generating violent waves of energy that compress yet more of the arms' dusty gas and accelerate the rate of star formation. Countless stars, those of lower mass that live long enough and do not explode, escape the arms and move into the Galaxy's general disk, where star formation can still take place, though at a slower pace. It is, of course, the mass tied up in all the stars that keeps the arms going, stars helping to beget more stars in a cycle within the cycle, the compression provided by stellar winds and explosions adding additional cycles within cycles.

Part of the Galaxy's mass is ionized: in a way that no one really understands, galactic rotation acts as a dynamo to produce a pervasive magnetic field that crudely tends to align itself with the spiral arms. Supernovae also launch fast-moving particles—cosmic rays—into the Galaxy's field, which then traps and bends their orbits through the interstellar matter. The cosmic rays penetrate dark clouds, where their ionizing energy aids processes of interstellar chemistry, a chemistry that, leading to yet unknown complexity, may help illuminate the origins of life. The cosmic rays also ionize the interiors of dense cores, allowing the Galaxy's magnetic field—the same field that traps the cosmic rays within the Galaxy—to latch on to the cores and slow their contraction and rotation. Stars are therefore the products not just of their local environments *but of the Galaxy as a whole.*

The young stars evolve, create new elements, add enriched mass back into the interstellar medium, and provide means of compression and cloud ionization for the creation of yet more stars. This picture of stellar recycling at first appears to be a monotonic progression. In fact, the grand loop, and all the intertwined subloops, consist of near-infinite sets of loops, all the action, star formation, and evolution *going on at the same time.*

Moreover, all these processes take place at different levels in different parts of the Galaxy because of different densities within the interstellar medium. The most tightly packed part of the Galaxy is the bulge, which contains enormous quantities of molecular gas. As a mixture of halo and disk—and a galactic feature with its own defining characteristics—our Galaxy's bulge contains stars with a huge range in metal content. Star formation proceeds at such a rate that some of its members are "super–metal-rich," with metal contents 10 times that of the Sun. Proceeding outward, the quantity of molecular gas declines precipitately at a distance of 2 kpc from the galactic center, then rises again between 4 and 6 kpc. Within this dramatic "molecular ring" are 2000 giant molecular clouds holding 2 billion solar masses of gas and dust, and in this region star formation again occurs frantically. Past the Sun, the number of clouds drops quickly to near zero 15 kpc away, where star formation consequently proceeds at a puny pace. Since the amount of matter ejected back into the Galaxy depends on the stellar death rate, which obviously is limited by the birthrate, the quantity of metals in the interstellar medium and in newly forming stars is much smaller at great distances from the galactic center. And all this space-dependent action is variable in time as well. There are cycles within cycles within cycles, all leading to

Paths' End

Sit alone upon an island shore
And watch the mating of the sand and sea
Encapsulate within your vision's core
The Universe's vast complexity.
Churning at the coast, the ocean hurls
A million sunlit bubbles to the sky,
Each flashing drop its own minuscule world,
Each a cosmos caught within your eye.
Now multiply this view around the Earth,
Then multiply afresh to space's end
Where heavens' stars began to give us birth,
Our Sun and Earth and selves a starry blend.
Only from such great infinity
Could all our hopes and dreams have come to be.

The paths of star formation and stellar evolution, winding in all their complexity through the Galaxy, have a countless number of endings; this is one of them.

We now know that our origins lie in the dust of interstellar space, that our Earth and our selves are condensates of the dark gaps between the stars, the same yawning expanses that are visible within the Milky Way on any clear night. Even with the enormous gaps in our knowledge that are yet to be bridged, we have at least solved a deep problem on a philosophical level. Continued inquiry, whether over years, decades, or centuries, should solve the myriad remaining questions in the flow of events that takes us from globules to planets, from the Galaxy to a field of summer flowers.

FURTHER READINGS

Aller, L. H. *Atoms, Stars, and Nebulae.* Cambridge, Mass.: Harvard University Press, 1971.

Aller, L. H. *Physics of Thermal Gaseous Nebulae.* Dordrecht: Reidel, 1984.

Beatty, J. K., and Chaikin, A., eds. *The New Solar System,* 3rd ed. Cambridge: Cambridge University Press, 1990.

Clark, D. H., and Stephenson, F. R. *The Historical Supernovae.* New York: Pergamon, 1977.

Cohen, M. *In Darkness Born: The Story of Star Formation.* Cambridge: Cambridge University Press, 1988.

Cowley, C. R. *An Introduction to Cosmochemistry.* Cambridge: Cambridge University Press, 1995.

Croswell, Ken. *The Alchemy of the Heavens.* New York: Anchor Books, 1995.

Dick, Steven J. *The Biological Universe,* Cambridge: Cambridge University Press, 1996.

Drake, F., and Sobel, D. *Is Anyone Out There? The Scientific Search for Extraterrestrial Intelligence.* New York: Delacorte, 1992.

Edmunds, M. G., and Terlevich, R., eds. *Elements and the Cosmos.* Cambridge: Cambridge University Press, 1992.

Goldsmith, D., and Owen, T. *The Search for Life in the Universe,* 2nd ed. Reading, Mass.: Addison-Wesley, 1992.

Kaler, J. B. *Stars.* Scientific American Library. New York: W. H. Freeman, 1992.

Kaler, J. B. *Stars and Their Spectra: An Introduction to the Spectral Sequence.* Cambridge: Cambridge University Press, 1989.

Levy, E. H., and Lunine, J. I., eds. *Protostars and Planets III.* Tucson: University of Arizona Press, 1993.

Marschall, L. A. *The Supernova Story.* New York: Plenum, 1988.

Morrison, D. *Exploring Planetary Worlds.* Scientific American Library. New York: W. H. Freeman, 1993.

Morrison, D., and Owen, T. *The Planetary System,* 2nd ed. Reading, Mass.: Addison-Wesley, 1996.

Norton, O. R. *Rocks from Space.* Missoula, Mont.: Mountain Press, 1994.

Osterbrock, D. E. *Astrophysics of Gaseous Nebulae,* 2nd ed. New York: W. H. Freeman, 1974.

Pottasch, S. R. *Planetary Nebulae.* Dordrecht: Reidel, 1984.

Verschuur, G. L. *Interstellar Matters.* New York: Springer-Verlag, 1989.

Whipple, F. L. *The Mystery of Comets.* Washington, D.C.: Smithsonian Institution Press, 1985.

Zuckerman, B., and Malkan, M. A. *The Origin and Evolution of the Universe.* Sudbury, Mass.: Jones & Bartlett, 1996.

Sources of Illustrations

Frontispiece by Tomo Narashima; *diagrams and graphs* by Fine Line Illustrations.

Front cover image: C. R. O'Dell and K. P. Handron (Rice University) and NASA.

Background image on chapter–opening pages: © Anglo-Australian Observatory.

Back cover image: © Anglo-Australian Observatory.

Chapter 1 *Facing page 1:* Hale Observatories. *Page 5* (left): National Optical Astronomy Observatories; (right): T. Boroson (NOAO/USGP), W. C. Keel (University of Alabama), and Kitt Peak National Observatory. *Page 6:* Dennis diCicco. *Page 7:* Royal Astronomical Society. *Page 8:* M. E. Killion. *Page 9:* © Anglo-Australian Observatory. *Page 11:* Palomar Observatory, California Institute of Technology. *Page 12: The Annals of the Harvard College Observatory,* vol. V, 1867, frontispiece. *Page 14:* Modified from Edward R. Harrison, *Cosmology* (Cambridge: Cambridge University Press, 1981), 272. *Page 15:* National Optical Astronomy Observatories. *Page 17:* From *The Scientific Papers of Sir William Huggins* (London: William Wesley and Sons, 1909). *Page 18* (left): From *The Scientific Papers of Sir William Huggins* (London: William Wesley and Sons, 1909), 110; (right): Kitt Peak National Observatory. *Page 20:* William Herschel, *Symphony in D Major,* ed. Sterling E. Murray, in *The Symphony 1720–1840,* ed. Barry S. Brook, series E, vol. 3 (New York: Garland, 1983), 3. *Page 22:* Harvard College Observatory.

Chapter 2 *Page 28:* C. R. O'Dell and S. K. Wong (Rice University) and NASA. *Page 31:* © Royal Observatory, Edinburgh/Anglo-Australian Observatory. *Page 32:* © Carnegie Institution of Washington. *Page 33:* Hui Yang (University of Illinois), J. Hester (University of Arizona), and NASA. *Page 36: The Facts on File Dictionary of Astronomy,* 2nd ed., ed. Valerie Illingworth (New York: Facts on File, 1985), 168. *Page 39:* © Anglo-Australian Observatory. *Page 43:* Lawrence H. Aller, Lick Observatory. *Page 47:* A. B. Wyse, Lick Observatory. *Page 53:* Palomar Observatory, California Institute of Technology. *Page 54:* Royal Observatory, Edinburgh.

Chapter 3 *Page 56:* © Anglo-Australian Observatory. *Page 58:* Gary Urton, in *Proceedings of the American Philosophical Society* 125 (1981), 110. *Page 59:* © Dirk Hoppe. *Page 60:* © Anglo-Australian Observatory. *Page 61* (top): Photo courtesy of Mrs. Joyce Bok Ambruster and Steward Observatory; (bottom): Mary Lea Shane Archives of the Lick Observatory. *Page 63* (left): Lick Observatory; (right): Data from M. Wolf, "Über den dunklen Nebel NGC 6960," *Astronomische Nachrichten* 219 (1923), 109–115. *Page 64:* Akira Fujii. *Page 65:* Lowell Observatory. *Page 66:* © IAC/RGO/Malin. *Page 68:* © Anglo-Australian Observatory. *Page 70:* © IAC/RGO/Malin. *Page 73:* Frederick Edwin Church, American, 1826–1900. *Twilight in the Wilderness,* 1860s. Oil on canvas, 101.6 × 162.6 cm. © The Cleveland Museum of Art, 1996, Mr. and Mrs. William H. Marlatt Fund, 1965.233. *Page 74:* Data from B. D. Savage and J. S. Mathis, "Observed Properties of Interstellar Dust," *Annual Review of Astronomy and Astrophysics* 17 (1979), 73–111. *Page 77:* NASA/Jet Propulsion Laboratory.

Chapter 4 *Page 80:* M. Normandeau, A. R. Taylor, and P. E. Dewdney, Dominion Radio Astrophysical Observatory, National Research Council of Canada. *Page 83* (left): Palomar Sky Survey; (right): Mount Wilson and Las Campanas Observatories. *Page 87:* Dap Hartmann and W. B. Burton, *Atlas of Galactic Neutral Hydrogen,* atlas and CD-ROM (Cambridge: Cambridge University Press, in press). *Page 88* (top): Data from NASA (*IUE*); (bottom): Data from E. B. Jenkins, "Element Abundances in the Interstellar Atomic Material," in *Interstellar Processes,* ed. D. J. Hollenbach and H. A. Thronson, Jr. (Dordrecht: Reidel, 1987), 533–559. *Page 89:* J. P. Bradley, S. A. Sanford, and R. M. Walker, "Interplanetary Dust Particles," in *Meterorites and the Early Solar System,* ed. J. F. Kerridge and M. S. Matthews (Tucson: University of Arizona Press, 1988), 861–895. *Page 90:* William Herschel, "On the Construction of the Heavens," *Philosophical Transactions* 75 (1785), 213–266. *Page 91:* Modified from Gareth Wynn-Williams, *The Fullness of Space* (New York: Cambridge University Press, 1992), 37. *Page 93:* © Anglo-Australian Observatory. *Page 94:* Courtesy of Gart Westerhout.

Chapter 5 *Page 102:* © Ken Edward/Science Source/Photo Researchers, Inc. *Page 108:* Yerkes Observatory. *Page 109:* Mount Wilson Observatory. *Page 112:* Lewis Snyder et al., Berkeley Illinois Maryland Association (BIMA). *Page 113:* Data from Gerrit L.Verschuur, "Interstellar Molecules," *Sky & Telecope* 83, no. 4 (April 1992), 379–384. *Page 118* (left): Data from F. Salama and L. J. Allamandola, "The Ultraviolet and Visible Spectrum of the Polycyclic Aromatic Hydrocarbon $C_{10}H_8{}^+$," *The Astrophysical Journal* 395, no. 1 (10 August 1992), 301–306; (right): Peter Jenniskens and NASA/Ames Research Center. *Page 119:* John Bally, University of Colorado. *Page 121:* Thomas Dame, Harvard University. *Page 122:* Modified from Nicholas Z. Scoville, "Interstellar Medium, Galactic Molecular Hydrogen," in *The Astronomy and Astrophysics Encyclopedia,* ed. Stephen P. Maran (New

York: Van Nostrand Reinhold, 1992), 373–375. *Page 124:* S. S. Prasad et al., "Chemical Evolution of Molecular Clouds," in *Interstellar Processes,* ed. D. J. Hollenbach and H. A. Thronson, Jr. (Dordrecht: Riedel, 1987), 631–667. *Page 125:* A. G. G. M. Tielens et al., "Anatomy of the Photodissociation Region in the Orion Nebula," *Science* 262 (1993), 86–89. *Page 126* (left): K. Sunada and Y. Kitamura, Nobeyama Radio Observatory; (right): Palomar Sky Survey.

Chapter 6 *Page 128:* J. Hester and P. Scowen (Arizona State University) and NASA. *Page 131:* Ronald A. Oriti. *Page 132:* Gary Goodman. *Page 133:* UCO/Lick Observatory Photo/Image. *Page 136:* Ian McLean/UCLA. *Page 137:* Data from G. Basri and C. Bertout, "T Tauri Stars and Their Accretion Disks," in *Protostars and Planets III* (Tucson: University of Arizona Press, 1993), 543–566. *Page 139:* J. Hester (Arizona State University), the WFPC 2 Investigation Definition Team, and NASA. *Page 140:* Annelia Sargent. *Page 141*: C. Burrows (STScI and ESA), the WFPC 2 Investigation Definition Team, and NASA. *Page 142:* Serge Koutchmy, Leon Golub, SAO, and CFH-T. *Page 144:* Modified from Y. Fukui et al., "Molecular Outflows," in *Protostars and Planets III* (Tucson: University of Arizona Press, 1993), 623–639. *Page 145:* (background photograph): Palomar Sky Survey; (contours): Modified from Ronald L. Snell, Robert B. Loren, and Richard L. Plambeck, "Observations of CO in L1551: Evidence for Stellar Wind Driven Shocks," *The Astrophysical Journal* 239, no. 1 (1 July 1980), L17–L22. *Page 146:* J. Morse (STScI) and NASA. *Page 147:* © Anglo-Australian Observatory. *Page 148:* Adapted from Ronald J. Maddalena et al., "The Large System of Molecular Clouds in Orion and Monoceros," *The Astrophysical Journal* 303 (1 April 1986), 375–391. *Pages 152 and 153:* Alan P. Boss, Carnegie Institution of Washington. *Page 154:* Modified from Steven W. Stahler, "Deuterium and the Stellar Birthline," *The Astrophysical Journal* 332 (1988), 804–822; tracks by Icko Iben, Jr.

Chapter 7 *Page 156:* Observatoire de Haute-Provence, France. *Page 159:* NASA. *Page 162:* Reta Beebe and Amy Simon (New Mexico State University) and NASA. *Pages 163 and 164:* NASA. *Page 165:* © 1996 Kurt Liffman. *Page 166:* NASA/Johnson Spaceflight Center. *Page 171:* David Jewitt and Jane Luu. *Page 173:* Virgil L. Sharpton, Lunar and Planetary Institute. *Page 174* (left): James V. Scotti, Kitt Peak National Observatory; (right): Richard P. Binzel, Massachusetts Institute of Technology. *Page 175:* M. J. McCaughrean (Max Planck Institute for Astronomy), C. R. O'Dell (Rice University), and NASA. *Page 176:* C. Burrows (STScI and ESA), J. Krist (STScI), the WFPC 2 Investigation Definition Team, and NASA. *Pages 178 and 179:* Geoffrey W. Marcy and Paul Butler, San Francisco State University Planet Search Project.

Chapter 8 *Page 184:* J. P. Harrington and K. J. Borkowski (University of Maryland) and NASA. *Page 186:* Yerkes Observatory. *Page 187:* National Optical Astronomy Observatories. *Page 189* (left): C. R. O'Dell and K. P. Handron (Rice University) and NASA; (right): © Anglo-Australian Observatory. *Page 190* (left): NOAO/George Jacoby; (right): S. Heap, NASA/Goddard Space Flight Center. *Page 191:* Bruce Balick, National Optical Astronomy Observatories. *Page 192:* G. H. Jacoby, National Optical Astronomy Observatories. *Page 194* (top): Bruce Balick, University of Washington; images obtained using the 2.1-m telescope at Kitt Peak National Observatory of the National Optical Astronomy Observatories, Tucson; (bottom): picture of NGC 7027 by James R. Graham, obtained with the 200-inch *Hale Telescope,* which first appeared in *The Astronomical Journal,* vol. 105, 1993. *Page 195:* Mount Wilson Observatory. *Page 197:* Adapted from James B. Kaler, *Stars and Their Spectra* (Cambridge: Cambridge University Press, 1989), 232. *Page 200:* Adapted from Icko Iben, Jr., "Single and Binary Star Evolution," *Astrophysical Journal Supplements* 76 (1991), 55-114. *Page 201:* John Bieging and Berkeley Illinois Maryland Association (BIMA). *Page 203:* R. Sahai and J. Trauger (JPL), the WFPC 2 Science Team, and NASA. *Page 204:* NOAO/George Jacoby. *Page 206:* Adapted from Bruce Balick, "The Shapes and Shaping of Planetary Nebulae," in *Planetary Nebulae,* Proceedings of the 131st Symposium of the International Astronomical Union (Boston and Dordrecht: Kluwer, 1989), 83–92. *Page 208:* Bruce Balick and Jason Alexander (University of Washington), Arsen Hajian (U.S. Naval Observatory), Yervant Terzian (Cornell University), Mario Perinotto and Patrizio Patriarchi (University of Florence), and NASA.

Chapter 9 *Page 210:* R. Chris Smith, University of Michigan and CTIO. *Page 213:* J. Morse (University of Colorado) and NASA. *Page 215:* Antonella Nota and Mark Clampin. *Page 216:* Karen Kwitter and You-Hua Chu. *Page 218:* A. Burrows, J. C. Hays, and B. Fryxell, University of Arizona. *Page 220:* © Anglo-Australian Observatory. *Page 223:* Courtesy of National Parks Service. *Page 225* (left): © 1994 Richard J. Wainscoat and John Kormendy, Institute for Astronomy, University of Hawaii; (right): Courtesy of Birr Castle Demesne. *Page 227:* Arecibo Radio Observatory, in Jeremiah P. Ostriker, "The Nature of Pulsars," *Scientific American* 224, no. 1 (January 1971), 48–60. *Page 228:* Kitt Peak National Observatory. *Page 229:* J. Hester and P. Scowen (Arizona State University) and NASA. *Page 230* (left): James R. Graham, Nancy A. Levenson, and McDonald Observatory; (right): James R. Graham, Nancy A. Levenson, and ROSAT. *Page 232:* Rosa Williams et al., CTIO.

Chapter 10 *Page 234:* R. Williams and the Hubble Deep Field Team (STScI) and NASA. *Page 242:* James B. Kaler.

INDEX

Selected Books in the Scientific American Library Series

POWERS OF TEN
by Philip and Phylis Morrison and the Office of Charles and Ray Eames

DRUGS AND THE BRAIN
by Solomon H. Snyder

EYE, BRAIN, AND VISION
by David H. Hubel

ANIMAL NAVIGATION
by Talbot H. Waterman

SLEEP
by J. Allan Hobson

FROM QUARKS TO THE COSMOS
by Leon M. Lederman and David N. Schramm

SEXUAL SELECTION
by James J. Gould and Carol Grant Gould

THE NEW ARCHAEOLOGY AND THE ANCIENT MAYA
by Jeremy A. Sabloff

A JOURNEY INTO GRAVITY AND SPACETIME
by John Archibald Wheeler

BEYOND THE THIRD DIMENSION
by Thomas F. Banchoff

THE SCIENCE OF WORDS
by George A. Miller

ATOMS, ELECTRONS, AND CHANGE
by P. W. Atkins

VIRUSES
by Arnold J. Levine

DIVERSITY AND THE TROPICAL RAINFOREST
by John Terborgh

STARS
by James B. Kaler

EXPLORING BIOMECHANICS
by R. McNeill Alexander

CHEMICAL COMMUNICATION
by William C. Agosta

GENES AND THE BIOLOGY OF CANCER
by Harold Varmus and Robert A. Weinberg

SUPERCOMPUTING AND THE TRANSFORMATION OF SCIENCE
by William J. Kaufmann III and Larry L. Smarr

MOLECULES AND MENTAL ILLNESS
by Samuel H. Barondes

EXPLORING PLANETARY WORLDS
by David Morrison

EARTHQUAKES AND GEOLOGICAL DISCOVERY
by Bruce A. Bolt

THE ORIGIN OF MODERN HUMANS
by Roger Lewin

THE EVOLVING COAST
by Richard A. Davis, Jr.

THE LIFE PROCESSES OF PLANTS
by Arthur W. Galston

IMAGES OF MIND
by Michael J. Posner and Marcus E. Raichle

THE ANIMAL MIND
by James L. Gould and Carol Grant Gould

MATHEMATICS: THE SCIENCE OF PATTERNS
by Keith Devlin

A SHORT HISTORY OF THE UNIVERSE
by Joseph Silk

THE EMERGENCE OF AGRICULTURE
by Bruce D. Smith

ATMOSPHERE, CLIMATE, AND CHANGE
by Thomas E. Graedel and Paul J. Crutzen

AGING: A NATURAL HISTORY
by Robert E. Ricklefs and Caleb Finch

INVESTIGATING DISEASE PATTERNS: THE SCIENCE OF EPIDEMIOLOGY
by Paul D. Stolley and Tamar Lasky

GRAVITY'S FATAL ATTRACTION: BLACK HOLES IN THE UNIVERSE
by Mitchell Begelman and Martin Rees

CONSERVATION AND BIODIVERSITY
by Andrew P. Dobson

PLANTS, PEOPLE, AND CULTURE: THE SCIENCE OF ETHNOBOTANY
by Michael J. Balick and Paul Alan Cox

LIFE AT SMALL SCALE: THE BEHAVIOR OF MICROBES
by David B. Dusenbery

PATTERNS IN EVOLUTION: THE NEW MOLECULAR VIEW
by Roger Lewin

CYCLES OF LIFE: CIVILIZATION AND THE BIOSPHERE
by Vaclav Smil